Landscape Maintenance Pest Control

PESTICIDE
APPLICATION
COMPENDIUM
7

Landscape Maintenance Pest Control

Patrick J. O'Connor-Marer
Pesticide Safety Training Coordinator
IPM Education and Publications
University of California, Davis

University of California
Statewide Integrated Pest Management Program
Agriculture and Natural Resources
Publication 3493

2006

ORDERING
For information about ordering this publication, contact:

University of California
Agriculture and Natural Resources
Communication Services—Publications
6701 San Pablo Avenue, 2nd Floor
Oakland, California 94608-1239

Telephone 1-800-944-8849
(510) 642-2431
FAX (510) 643-5470
E-mail: danrcs@ucdavis.edu
Visit the ANR Communication Services Web site at danrcs.ucdavis.edu.

Publication 3493

UC PEER REVIEWED This publication has been anonymously peer reviewed for technical accuracy by University of California scientists and other qualified professionals. This review process was managed by the ANR Associate Editor for Pest Management.

ISBN-13: 978-1-879906-71-6
ISBN-10: 1-879906-71-6
Library of Congress Control Number: 2005929433

Funding for this manual was provided in part by the Extension Service, U.S. Department of Agriculture, under special project Section 3(d), Pesticide Safety Education Program, and through a National Institute for Occupational Safety and Health cooperative agreement number U 07/CCU 9066162-03 with the Western Center for Agricultural Health and Safety at UC Davis.

Acknowledgments

This manual was produced under the auspices of the University of California Statewide Integrated Pest Management Program through a Memorandum of Understanding between the University of California and the California Department of Pesticide Regulation. It was prepared under the direction of Richard T. Roush, Director, UC Statewide IPM Program, University of California, Davis.

Research and Writing Assistance

Diane Clarke performed extensive research, prepared drafts of several chapters, and coordinated the Expert Panel. Melanie Zavala developed the initial outline, coordinated advisory committee input, and arranged for many of the photographs.

Production

Design: Seventeenth Street Studios
Production Coordination: Pauli Galin
Page Composition: Valerie Griffith
Photographs: Jack Kelly Clark and Suzanne Paisley
Drawings: David Kidd, Valerie Griffith, Celeste Rusconi
Editing: Jim Coats
Cover Design: Valerie Griffith
Technical Editor: Mary Louise Flint

Technical Committee, Contributors, and Principal Reviewers

The following people provided ideas, information, and suggestions and reviewed the many manuscript drafts. Some served on an advisory committee or the Expert Panel; others were key resources and reviewers in specific subject matter areas.

M. Barneveld, Valley Crest, San Francisco
C. Clifton, Lawnscape Systems, Inc., Montclair
L. Costello, University of California Cooperative Extension, Half Moon Bay
J. Downer, University of California Cooperative Extension, Ventura
S. Dreistadt, University of California Statewide IPM Program, Davis
M. L. Flint, University of California Statewide IPM Program, Davis
G. Forth, State Center Community College District, Fresno
G. Gallardo, Valley Crest, San Francisco

A. Gallo, California Department of Pesticide Regulation, Sacramento
B. Gardner, Hayward Area Recreation and Parks District, Hayward
P. Geisel, University of California Cooperative Extension, Fresno
A. Harivandi, University of California Cooperative Extension, Alameda
J. Haslett, Innovative Plantscapes, Inc., Chatsworth
R. Jones, City of Davis Parks and Recreation Department, Davis
A. King, University of California Cooperative Extension, San Mateo–San Francisco
J. Letterman, Pesticide Applicator's Professional Association, Salinas
T. LoCoco, City of Santa Rosa Recreation and Parks Department, Santa Rosa
J. Newman, University of California Cooperative Extension, Ventura
E. Perry, University of California Cooperative Extension, Modesto
J. Racklin, John Racklin Weed and Pest Control, Fresno
M. Takeda, California Department of Pesticide Regulation, Sacramento
B. Thompson, Pest Options, Inc., Orange
C. Wilen, University of California Statewide IPM Program, San Diego
S. Youney, California State University Sonoma, Landscape Department, Santa Rosa

Contents

Tables

Sidebars

Introduction

THE PROPER HANDLING of pesticides in turf, landscape, and interiorscape situations involves many special skills and responsibilities. It is important that you recognize the hazards associated with pesticides and that you know how to avoid them. Furthermore, you must be familiar with all applicable local, state, and federal laws regulating the use, storage, transportation, application, and disposal of the pesticides you are using. In addition, if you supervise pesticide handlers you are responsible for seeing that these employees are trained in the safe and proper handling and use of the pesticides.

This book is one volume of the University of California's Pesticide Application Compendium. Our purpose in publishing the book is to help you learn safe and effective ways of using pesticides on turf, landscapes, and interiorscapes. It also tells you how to prevent accidents and avoid injuries and environmental problems.

How to Use this Book

Use this book as a study guide as you prepare for the California Qualified Applicator Certificate (QAC) or Qualified Applicator License (QAL) examinations in the Landscape Maintenance Pest Control and Maintenance Gardener categories. The California Department of Pesticide Regulation (DPR) uses these tests to certify individuals, pest control businesses, commercial applicators, and landscape maintenance personnel who apply pesticides as part of their work.

This book's chapters cover the information you need to know in order to properly use pesticides on landscapes, interiorscapes, and turfgrass. Tables and sidebars provide additional information that is important for you to know. Specialized words and terms are defined in the Glossary at the back of the book.

What You Need to Know

The DPR exams assess your competence in handling or supervising the handling of pesticides. Questions on these exams are similar to the review questions at the end of each chapter of this book. The exams cover several areas of skill to make sure you have met DPR's knowledge expectations. Knowledge expectations for landscape maintenance pest managers and maintenance gardeners are listed in Sidebar 1. The following paragraphs further describe what you need to know and where you can find relevant information in this book.

Pests and Pest Management. You must be able to recognize and identify common pests and the signs of their presence, including the damage they cause. You should also understand the concepts of pest management, including principles of integrated pest management (IPM). (See Chapters 1 and 2.)

The Pesticide Label. You must demonstrate that you can read and understand pesticide labels and associated labeling information. This includes recognizing restrictions on the use of certain pesticides, understanding worker safety and protection requirements, and recognizing and understanding how to protect environmentally sensitive areas. (See Chapter 3.)

Pesticide Types and Formulations. Before you can select the right pesticide to manage a landscape, turf, or interiorscape pest, you need to know how the various available chemicals work (indicated by their "type") and what formulations are available for those pesticides. Different pesticide formulations and types react differently with target pests, nontarget organisms, plants, and the environment. (See Chapter 3.)

Laws and Regulations Affecting Pesticide Use. You need to be familiar with current pesticide regulations and you must know how to locate regulatory information. (See Chapter 4 and the *Laws and Regulations Study Guide* available from DPR.)

Protecting the Environment. You must recognize how pesticides can harm nontarget organisms. You need to understand how environmental contamination occurs and how you can prevent it. You need to know how to properly handle, store, and dispose of leftover pesticides, pesticide mixtures, and empty containers. (See Chapters 5 and 6.)

Recognizing and Preventing Pesticide Poisoning. You must recognize common symptoms of pesticide poisoning and demonstrate that you know how to protect yourself and others from exposure through the use of personal protective equipment (PPE). You also need to understand first aid procedures for pesticide exposure. (See Chapters 5, 6, and 7.)

Pesticide Emergencies. You must know where to get first aid information for pesticide exposure victims and you must know how to administer first aid to an exposure victim. You must understand the procedures for dealing with pesticide spills and fires. You have to be able to recognize the problems associated with the misapplication of pesticides and know what to do if such a thing occurs. (See Chapter 7.)

Using Pesticides Effectively. Safety is a key element in any pest control operation, but your overall goal is to control the target pest, so the operation must also be effective and economical. Application timing and equipment and the chemical or technique you choose to use are just a few of the factors that we explain. (See Chapter 8.)

Mixing and Applying Pesticides. You need to know how to prepare pesticides in the proper concentrations for use under various circumstances. You must also be able to select the appropriate application equipment and correctly calibrate that equipment. You need to demonstrate your ability to prevent off–site movement of pesticides during application. (See Chapters 9, 10, and 11.)

Using the Review Questions

At the end of each chapter are several review questions to test your grasp of the information presented in that chapter. These questions are written in the same format as the questions on the DPR examinations.

Begin your study by reading a chapter, and then read the review questions at the end of that chapter. Make a note of any subject material you do not fully understand. Then review the chapter to locate the sections that deal with that information and study those sections again.

When you finish studying the chapter, answer each of the review questions. Check your answers against the correct answers at the end of the book. If you missed any of the questions, go back and reread the sections of the chapter that cover that information.

SIDEBAR 1

Knowledge Expectations: Examinations for the Qualified Applicator Certificate and License, Landscape Maintenance Pest Control and Maintenance Gardener Categories

PEST IDENTIFICATION

- ☐ must have a general understanding of pest identification
- ☐ must be able to distinguish damage caused by pests with sucking mouthparts from that caused by pests with chewing mouthparts
- ☐ must understand the concept of pest life cycles and their importance to effective pest identification and management
- ☐ must be able to tell the difference between broadleaf and narrow-leaf (grassy) weeds
- ☐ must know that plants can be injured by inadequate or improper cultural practices, insects and mites, vertebrates, weeds, fungi, bacteria, viruses, nematodes, environmental conditions such as drought and nutritional deficiency, and chemicals such as air pollutants and spray drift
- ☐ must know where to access pest identification information and resources

PEST MANAGEMENT/USING PESTICIDES EFFECTIVELY

- ☐ must understand general principles of pest management, including integrated pest management (IPM)
- ☐ must know how to protect beneficial organisms
- ☐ must be able to recognize plant stress and have a general awareness of the range of potential causes
- ☐ must understand what natural enemies are and how they control pests
- ☐ must understand the importance of monitoring weather conditions prior to and during an outdoor pesticide application
- ☐ must understand that there is a range of pest control methods to choose from, including cultural, chemical, and biological forms
- ☐ must understand pesticide selectivity
- ☐ must understand the principles of the selective use of pesticides
- ☐ must understand the benefits of using pest-resistant plant varieties
- ☐ must be able to recognize when pest control efforts have not been successful
- ☐ must know where to access pest management information, including on the internet

PESTICIDES

- ☐ must have a thorough understanding of what a pesticide is
- ☐ must understand the concept of different toxicities of pesticides
- ☐ must be familiar with the different pesticide chemical groups
- ☐ must understand the concept of a pesticide's "mode of action"
- ☐ must recognize factors that influence the efficacy of pesticides
- ☐ must be familiar with the different types of pesticide formulations
- ☐ must understand what adjuvants are for
- ☐ must understand how to recognize incompatibility in pesticide mixtures
- ☐ must understand ways to reduce incompatibility problems

PESTICIDE LAWS AND REGULATIONS

- ☐ must have a general understanding of pesticide laws and regulations
- ☐ must understand the various parts of a pesticide label and be able to find specific information on the label
- ☐ must recognize what is and is not an acceptable deviation from pesticide label requirements
- ☐ must understand how to avoid using pesticides in conflict with the label instructions
- ☐ must understand the importance of only using products that are registered for use on the application site
- ☐ must understand the proper and legal disposal of excess materials and empty pesticide containers
- ☐ must know how to properly transport pesticides
- ☐ must understand what "service containers" are and how these must be labeled
- ☐ must know how to properly store pesticide containers and equipment
- ☐ must understand that pesticides must be stored in proper containers and never in food storage containers (e.g., soft drink bottles)
- ☐ must understand worker safety requirements (especially training and personal protective equipment (PPE)

HAZARDS ASSOCIATED WITH PESTICIDE USE

- ☐ must have an understanding of pesticide hazards
- ☐ must understand how people get exposed to pesticides in landscape, turfgrass, and interiorscape settings
- ☐ must understand how pesticides enter the body
- ☐ must understand the health effects of exposure to pesticides
- ☐ must understand how pesticides get into surface water and groundwater
- ☐ must understand the impact of pesticides on nontarget organisms
- ☐ must understand the causes of phytotoxicity in landscape, interiorscape, and turfgrass applications
- ☐ must understand that spray drift of soil-applied herbicides can injure sensitive plants
- ☐ must understand that pesticide persistence in the soil is one cause of phytotoxicity
- ☐ must know how to recognize the results of phytotoxicity (including abnormal growth, leaf drop, and discolored, curled, and spotted leaves)
- ☐ must understand that when plants are young and tender they are more susceptible to phytotoxicity
- ☐ must understand how to consult the label about avoiding phytotoxicity for specific plants or varieties

PROTECTING PEOPLE AND THE ENVIRONMENT

- ☐ must have a general understanding of how to protect people and the environment
- ☐ must ensure that health and safety hazards have been considered before a pesticide is applied
- ☐ must know how to prevent humans, pets and other domestic animals, and ornamental fish from contacting hazardous amounts of pesticides within the treated area
- ☐ must recognize hazards at the application site that could endanger people or the environment during a pesticide application
- ☐ must know how to handle pesticides safely within landscapes, interiorscapes, and turfgrass environments before, during, and after spray applications
- ☐ must recognize the ways pesticides can damage surfaces of objects in areas where pesticides are applied
- ☐ must understand how to prevent hazardous amounts of pesticide from drifting into nontarget areas
- ☐ must know how to select, fit, care for, and use personal protective equipment
- ☐ must understand how to recognize, treat, and prevent heat-related illness that is sometimes associated with the use of personal protective equipment
- ☐ must understand how to reduce the environmental impact of pesticides
- ☐ must understand how to safely clean application equipment and dispose of empty containers
- ☐ must understand proper personal hygiene practices to use after handling pesticides
- ☐ must understand the legal liabilities associated with pesticide application
- ☐ must read the label and be familiar with the characteristics of each pesticide that may be applied in landscapes or interiorscapes or on turfgrass
- ☐ must be aware of the ultimate fate of pesticides in the environment through drift, volatilization, photodecomposition, runoff, leaching, degradation by microorganisms, dilution by soil water, and attachment to soil particles
- ☐ must understand posting requirements
- ☐ must communicate with others who perform work in the treated area (e.g., mowers)

PESTICIDE EMERGENCIES

- ☐ must have a general understanding of how to deal with pesticide emergencies
- ☐ must know proper first aid procedures for pesticide exposure
- ☐ must know procedures for dealing with pesticide leaks and spills
- ☐ must understand how to deal with pesticide fires
- ☐ must know procedures for dealing with pesticide misapplications
- ☐ must know how to deal properly with mixing and loading errors
- ☐ must understand who to call in emergency situations

PESTICIDE APPLICATION EQUIPMENT

- ☐ must understand proper pesticide application procedures
- ☐ must be aware of the different types of landscape, interiorscape, and turfgrass pesticide application equipment
- ☐ must understand the components of liquid application equipment and the roles of each component
- ☐ must understand how to use bait and bait stations for vertebrate pests
- ☐ must know how to service and maintain pesticide application equipment
- ☐ must know how to monitor equipment during an application in order to spot problems as they arise
- ☐ must understand how different nozzle patterns affect application
- ☐ must understand the affect of sprayer pressure, nozzle type, and nozzle height on drift
- ☐ must understand how to properly clean tools and equipment

CALIBRATING PESTICIDE APPLICATION EQUIPMENT

- ☐ must have a general understanding of calibration
- ☐ must understand the reasons for calibrating pesticide application equipment
- ☐ must know how to calibrate for interiorscape, turfgrass, and ornamental applications
- ☐ must be able to calculate the amount of pesticide to apply to a known area
- ☐ must understand proper pesticide mixing procedures
- ☐ must know how to make a pesticide mixture that contains a certain percentage of active ingredient
- ☐ must know how to mix pesticides on a parts-per-million (ppm) basis
- ☐ must understand the importance of using the proper equipment for each job

1 Pest Identification

PESTS IN THE LANDSCAPE, on turf, and in interiorscape plantings cause problems in a variety of ways. The pests may be weeds that either compete with desirable plants for water, nutrients, and light or reduce the visual beauty of the landscape. Weeds can also interfere with the use of sports fields and may create hazards that could cause injury to people using these areas. Besides weeds, the pests may be insects, mites, fungi, bacteria, viruses, or nematodes that injure or disfigure plants or cause plant diseases, or they may be vertebrates such as rodents that make burrows and mounds and chew on plant roots, bark, and leaves (Figure 1-1).

Identifying a pest is an important first step toward managing it. First, you need to be certain that the damage you see is actually caused by a pest and not some other factor. In some cases the first pest you identify may be benign, and may not be causing damage at all. Even if an organism looks like a pest, it may not be the actual cause of damage or it may be causing so little damage that it does not warrant control.

Once you know the name of the pest, you can turn to publications and other resources that will tell you about its potential for damage and its life cycle, growth, and reproductive habits. With this information, you can more easily select effective management methods.

Pest misidentification or lack of information about an accurately identified pest could prompt you to choose the wrong management method or to apply the method at the wrong time. These are the two most frequent causes of pest management failure.

This chapter will give you some ways to identify pests. Four main groups of pests are:

- weeds (undesired plants)
- invertebrates (insects, spiders, mites, nematodes, snails, and slugs)
- vertebrates (rodents, other mammals, and birds)
- plant pathogens (bacteria, viruses, fungi, and other microorganisms)

For more information about pest identification, see Chapter 1 in *The Safe and Effective Use of Pesticides, Second Edition* (UC ANR Publication 3324). For help in identifying and managing specific turf and landscape plant pests, visit the "landscape, gardens, and turf" section of the UC Integrated Pest Management Program's Web site (http://www.ipm.ucdavis.edu) or refer to *Pests of Landscape Trees and Shrubs: An Integrated Pest Management Guide, Second Edition* (UC ANR Publication 3359).

You must also know the identity of the plants that you are protecting from pest damage (Figure 1-2). This information may be very helpful as you try to identify pests and may be necessary to help you select the proper pesticide. Proper identification of the plants can also help you avoid injury to landscape, turf, or interiorscape plants when you use pesticides. Many plant nurseries can assist you with plant identification.

WAYS TO IDENTIFY LANDSCAPE, TURF, AND INTERIORSCAPE PESTS

Identify pests by using the guidelines in this chapter, visiting the UC IPM

FIGURE 1-1.

Pests of several different types cause problems in landscape, turf, and interiorscape plantings. Shown here are examples of weeds (A), mites (B), insects (C), and plant diseases (D). Vertebrate animals such as gophers, moles, rats, mice, and squirrels are also occasional pests.

Web site, and consulting books such as those listed above. *Pests of Landscape Trees and Shrubs, Second Edition* includes useful host-pest diagnostic tables that can help you identify plant pests (Figure 1-3). Similar tables for turfgrass pests are in the online *UC Guide to Healthy Lawns* (http://www.ipm.ucdavis.edu/TOOLS/TURF/). You can also have the pest examined and identified by a specialist (see below).

The main difficulty in identifying certain insects and most mites, nematodes, and plant pathogens is their small size. Identification may require the use of a hand lens or microscope, special tests, or careful analysis of damage. This is why it may be important to know the type of plant the pest is attacking and its habitual location on the plant. Environmental conditions where you collect pests and the time of year of collection also may provide clues to the pest's identity.

Pests may have different physical forms depending on the stage in their life cycle and the time of year. Weed seedlings do not resemble mature weed plants. Many insect species undergo extreme changes in appearance as they develop from eggs through larval, pupal, and adult stages.

Identification Experts

Trained experts using special techniques and equipment can positively identify many pests such as nematodes, mites, insects, and plant pathogens. UC

FIGURE 1-2.

To aid in identifying pests, it is important to know the species of plant being attacked. This is a photograph of Vinca major.

FIGURE 1-3.

Pests of Landscape Trees and Shrubs, Second Edition (UC ANR Publication 3359) contains valuable diagnostic tables to help you identify plant pests.

Cooperative Extension environmental horticulture advisors have expertise in pest identification and they are in close contact with other University of California experts. County agricultural commissioners and their staff are also helpful. In addition, the California Department of Food and Agriculture maintains a pest identification laboratory.

When you are getting samples identified, be sure to keep the plant material fresh and undamaged. Provide complete information on where you found the pest and, if appropriate, include examples of pest damage. Weeds, disease organisms, and insects and mites all require different types of sampling and handling. Sidebars in the following sections provide suggestions on how to collect and prepare pest specimens for shipment. Be sure to tell the identification expert which plant species the pest was attacking.

Photographs and Drawings

Photographs and drawings often provide good visual descriptions of a pest and its damage (Figure 1-4). Some photos or drawings show you where to find unique or distinguishing external structures or body parts of the pest. For help in identification, use publications such as the California Department of Food and Agriculture's *Vertebrate Pest Control Handbook* and the University of California's *Pests of Landscape Trees and Shrubs, Second Edition* (UC ANR Publication 3359), *Pests of the Garden and Small Farm* (UC ANR Publication 3332), *Wildlife Pest Control around Gardens and Homes* (UC ANR Publication 21385), and various UC IPM Pest Notes available on the internet (http://www.ipm.ucdavis.edu/).

Characteristic Signs

Pests may leave signs of their presence that will help you identify them. For example,

- birds and rodents often build distinctive nests
- certain insects leave easily recognizable feeding damage
- mites may produce webbing on leaf surfaces
- rodents often leave identifying gnaw marks on tree trunks or other objects
- trails in grass or tracks in dirt

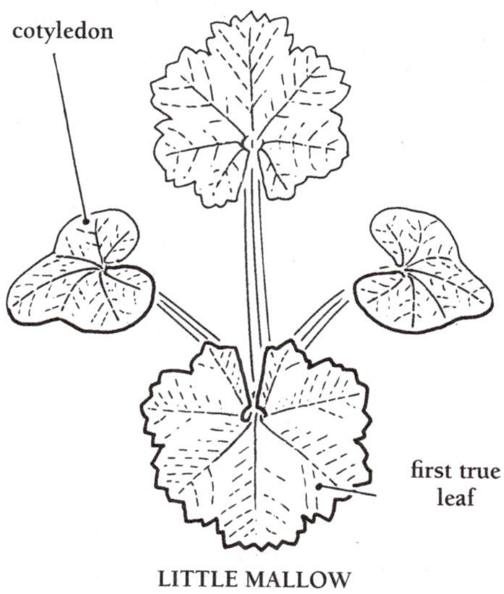

FIGURE 1-4.

Photographs and drawings are good tools to use when you are identifying a pest. This illustration shows the common weed little mallow, Malva parviflora.

sometimes help point to a pest's identity

- rodent fecal pellets and insect frass are distinctive and may assist in identification
- weeds may have unique flowers, seeds, fruits, or unusual growth habits
- fungi and other pathogens sometimes cause specific types of damage, deformation, or color changes to infected plants, and these clues can help you identify these organisms

WEEDS

Weeds are a nuisance because of their growth habit and size. They may detract from the beauty of the landscape, and some weeds create hazards. Even though some weeds are no more than an aesthetic problem, that alone is significant in many landscapes. Table 1-1 lists common weeds found in turf or landscaped areas.

Weeds adapt well to local climates, soils, and other external conditions. They compete successfully with cultivated plants for available resources. Most produce large quantities of seeds, even under adverse conditions. Seeds of some weeds remain dormant in the soil for extended periods, up to 20 or more years, before they germinate. Some weed seeds or fruits are specially adapted in ways that promote dispersal (Figure 1-5).

Many weeds are capable of reproducing through vegetative structures, such as stolons, rhizomes, and tubers. Cultural activities such as hoeing or mowing that break up and disperse these structures can actually end up producing more new weed plants (Figure 1-6). This is just one factor that makes many weeds persistent and difficult to eliminate.

Some weeds are native plants, but most are introduced species imported inadvertently or intentionally from other parts of the world. In their new location, some introduced plants become serious pests. This is because they have characteristics that allow them to establish easily and they can thrive under a wide range of conditions. Additionally, natural enemies or

TABLE 1-1.

*Common Weeds Found in Turf and Landscape Areas.**

Annual bluegrass
Bermuda buttercup
Bermudagrass
Clovers
Common knotweed
Common purslane
Crabgrass
Creeping woodsorrel
Dallisgrass
Dandelions
Dodder
Field bindweed
Green kyllinga
Kikuyugrass
Mistletoe
Nutsedge
Plantains
Poison oak
Russian thistle
Spotted spurge
Wild blackberries
Yellow starthistle

* See the UC IPM Program Web site at www.ipm.ucdavis.edu/PMG/selectnewpest.landscape.html#WEED for photos and management guidelines for common landscape weeds.

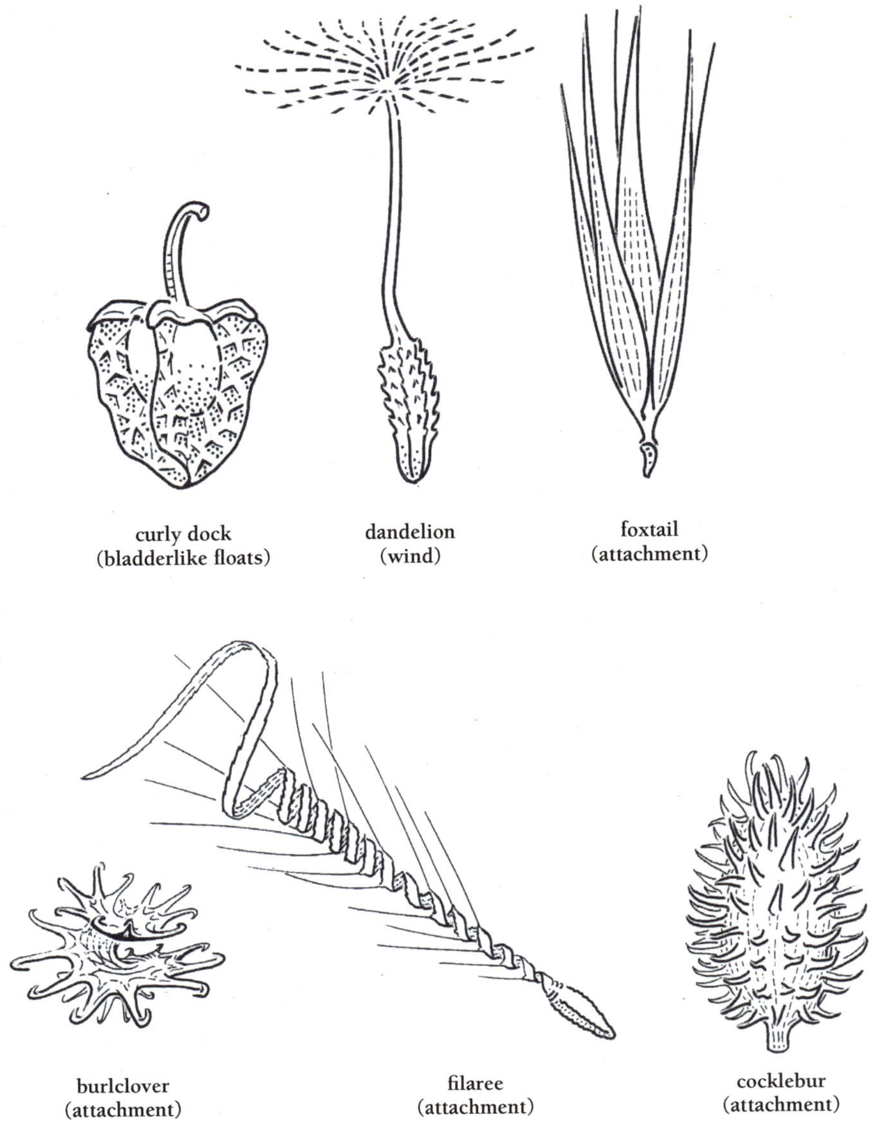

FIGURE 1-5.

As shown here, weed seeds have many adaptations that enable them to disperse. These special characteristics help the weeds compete with other plants.

diseases that suppress these plants in their native areas may not be present in their new location.

Identifying Weeds

A simple way to identify common weeds is to compare specimens with drawings and photographs. To do this, refer to publications such as those listed above as well as *Weeds of the West* (UC ANR Publication 3350) and the two of the UC IPM Pest Notes, *Weed Management in Lawns* and *Weed Management in Landscape* (http://www.ipm.ucdavis.edu). If you are unable to identify them from these sources, you probably need to send samples of the weed to an expert for identification. See Sidebar 2 for information on sampling and sending weeds for identification. Weed management is more effective when the weed plants are in their seedling stage, so begin sampling and identification early.

Specialists identify weeds by examining their flowers, leaves, stems,

FIGURE 1-6.

Cultural activities, such as mowing or dethatching, can spread some weeds to new areas. Seeds or other reproductive plant parts can move to other areas on mowing equipment.

and roots. Fruit, seeds, and special structures such as tubers or rhizomes are also useful for their identification characteristics, as are descriptions of the plants' growth habits.

Weed Classification. Most weeds fall within two major plant groups, the dicots and the monocots. Dicots, also called broadleaves, are plants that produce two seedling leaves (cotyledons).

SIDEBAR 2

Guidelines for Sampling and Sending Weeds for Weed Identification

SAMPLING

- ☐ Choose several plants that represent the species.
- ☐ Include stems, leaves, flowers (if present), and roots.
- ☐ Dig up weeds to prevent damage to roots.
- ☐ Shake plants lightly after digging to remove excess soil.

PREPARATION

- ☐ Keep plants chilled in an ice chest while you are in the field. If they cannot be shipped immediately, store them in a refrigerator.
- ☐ Place plants in plastic bags without moisture or press them between sheets of absorbent paper and encase these in heavy cardboard for protection.

LABELING

Attach a label to the outside of each sample. Include the following information on labels:

- ☐ Your name, address, and telephone number.
- ☐ Date samples were taken.
- ☐ Whether plants in question are annuals or perennials.
- ☐ Description of specific characteristics of the site where the weeds were growing.
- ☐ Any other information that would help in the identification of the weeds.
- ☐ Location where samples were taken, including names of nearby crossroads.

SHIPPING

- ☐ Contact the person or laboratory that will receive samples to determine the best method of shipping and to inform them that samples will be arriving.
- ☐ Pack samples in a sturdy, well-insulated container to prevent crushing or heat damage.
- ☐ Mark the package clearly and request that the shipper keep it in a cool location.
- ☐ Ship packages early in the week so they will arrive before a weekend.

The true leaves of dicots usually have net-like veins. Dicots generally grow as herbaceous plants (leafy and herb-like) or as woody plants (shrub- or tree-like). Parasitic dicots such as mistletoe and dodder are unique in their growth habits and characteristics: they take nutrients and water from host plants rather than from the soil.

Monocots produce only a single grasslike leaf in the seedling. Leaves of these plants typically have veins that run parallel to their long axes. Grasses, sedges, and rushes are monocots.

Mosses and liverworts belong to a unique group of plants known as bryophytes. Mosses differ from other types of plants in that they lack a vascular system to carry nutrients and water to various parts of the plant. They occasionally become pests in moist settings and sometimes cause problems on indoor or outdoor plants.

Algae are nonflowering primitive aquatic plants that can clog ornamental ponds, pools, streams, lakes, and drainage ditches. Algae are usually grouped with weed pests. Like more highly developed plants, algae carry out photosynthesis, a chemical process that converts light into energy. However, algae lack true stems, leaves, and flowers. They reproduce either through cell division or by producing spores.

FIGURE 1-7.

Plant seedlings, such as that of common lambsquarters shown here (top), differ in appearance from mature plants (bottom).

Weed Development. Most weeds pass through several growth stages, beginning with the seed. Most sprouted seeds (seedlings) are tender and vulnerable. They often are quite susceptible to herbicides. Seedlings differ in appearance from mature plants. This may make weeds difficult to identify (Figure 1-7), but because they are most easily controlled at the seedling stage it is important that you be able to identify them.

From seedlings, weeds continue to grow and produce foliage until they attain their mature size. They then enter a reproductive period and divert most of their energy into flowering and seed production. Weeds are mature once they have formed seeds. Annual plants (see below) continue to repeat these cycles of seedling, vegetative growth, and reproduction each year.

To manage weeds effectively, you will need to recognize the different growth stages. An understanding of these growth stages will make you better able to select an appropriate herbicide (and herbicide application rate) or other management method.

Weed Life Cycles. Weeds have annual, biennial, or perennial life cycles. Occasionally, a weed with one type of life cycle may behave as if it had a different life cycle. This may be attributable to weather or other influences. Milder temperatures, for example, can promote longer life cycles.

You can use the stage of the weed in its life cycle as a guide for proper timing of herbicide applications and other management activities. For instance, most perennial weeds are less susceptible to herbicides during their early bloom stage.

Annual plants live one year or less. Over the course of this period, they sprout from seeds, mature, and produce seeds for the next generation. Annual weeds are either summer annuals or winter annuals (Figure 1-8). Seeds of summer annuals sprout in the spring and the mature plants produce seeds and die during the late summer or fall. Some common summer annual weeds are spotted spurge, sowthistle, and common purslane.

Winter annuals sprout in the fall and grow over the winter. These produce seeds in the spring and then die, usually before summer. Mustard, wild oat, annual bluegrass, mallow, and burclover are examples of winter annual weeds found in outdoor landscapes (Figure 1-9).

Biennial plants usually live for two growing seasons, but in some areas such as the mild coastal regions of California biennial plants may live longer than this. These plants sprout and

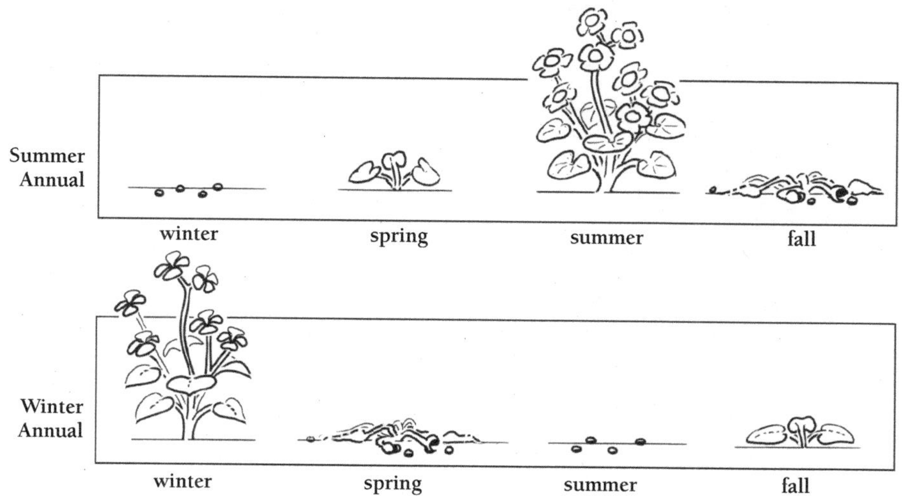

FIGURE 1-8.

Summer annual weeds germinate in spring and go to seed in late summer. Late winter annual weeds germinate in fall and produce seeds in late winter.

undergo vegetative growth during the first season. Then they flower, produce seeds, and die the following season. Bristly oxtongue is one biennial weed found in landscapes (Figure 1-10). Some species of perennial plants will go on living indefinitely.

Many perennials, such as certain species of trees, become dormant and lose their leaves each winter. Herbaceous perennials die back entirely during the winter and regrow each spring from roots or underground storage organs such as tubers, bulbs, or rhizomes. These storage organs also serve as a way for a number of major perennial weeds to reproduce and disperse. Examples of perennial weeds in turf and

FIGURE 1-9.

The weeds shown at right are examples of winter annual weeds typically found in landscapes. These include mustard (A), wild oat (B), annual bluegrass (C), and little mallow (D).

FIGURE 1-10.

Bristly oxtongue is one of the biennial weeds found in landscapes.

landscapes are oxalis, field bindweed, dandelion, nutsedge, poison oak, and bermudagrass (Figure 1-11). Of all weeds, perennial weeds are the most difficult to control.

Physical Features of Weeds. Weeds and most other plants have unique physical features that experts use as a basis for identification: features such as flowers, leaves, stems, special rooting structures, roots, fruits, and seeds.

Flowers. Flowers contain the weeds' sexual reproductive organs and differ widely between species. To use flowers as an identification aid, become familiar with the different flower parts (Figure 1-12). Flowers occur singly or as compound inflorescences.

An inflorescence is a group of flowers arising from a common main stem. Different names such as panicle, raceme, spike, head, umbel, and catkin describe the arrangement of flowers in an inflorescence (Figure 1-13).

Leaves. The arrangement and shape of leaves, vein patterns, and presence of

(A) (B) (C) (D) (E) (F)

FIGURE 1-11.

Examples of perennial weeds in turf and landscapes include Bermuda buttercup (A), oxalis (B), field bindweed (C), dandelion (D), nutsedge (E), and bermudagrass (F).

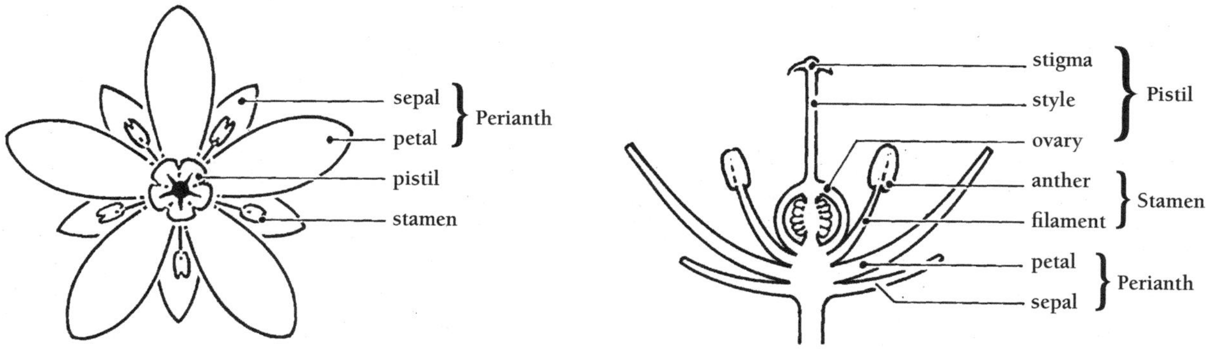

FIGURE 1-12.

Principle parts of a weed's flower. Many weeds can be correctly identified based on a clear description of their flowers.

spines or hairs are helpful identifying characteristics. When you are identifying weed seedlings, leaves are the only easily visible structures. Cotyledons and in some cases the first true leaves of a broadleaved plant differ from the plant's mature foliage. For grass identification, experts rely on the collar region of the plant where leaves separate from the main stem (Figure 1-14).

Stems. Stems form a plant's framework or "skeleton." They connect roots to other structures, providing structural support for leaves and flowers as well as channels for the transport of nutrients and water. Specialized stem modifications enable some weeds to reproduce vegetatively (Figure 1-15). Rhizomes are elongated underground stems that grow horizontally from the

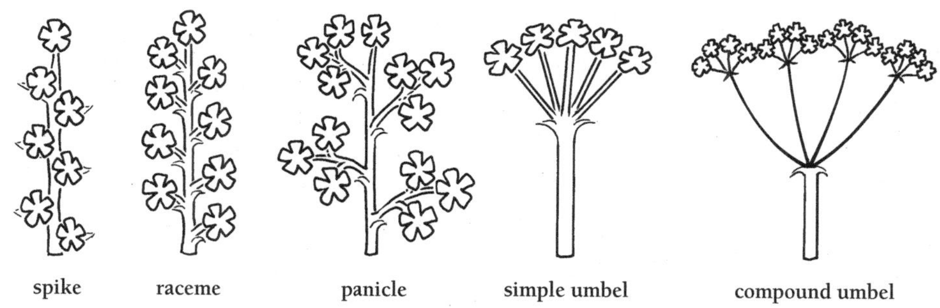

FIGURE 1-13.

As shown here, there are several types of inflorescences. These flower characteristics are very useful in weed identification.

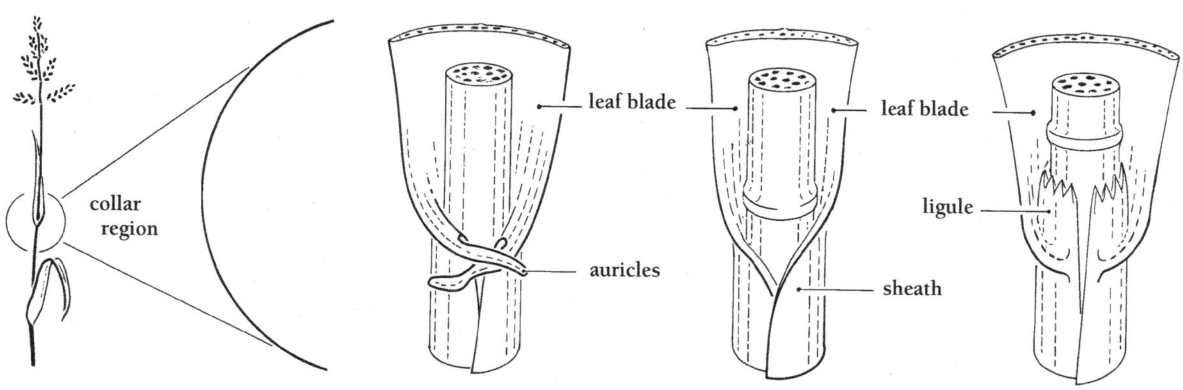

FIGURE 1-14.

In the collar region of a grass leaf are unique structures that are very important in identifying grass species.

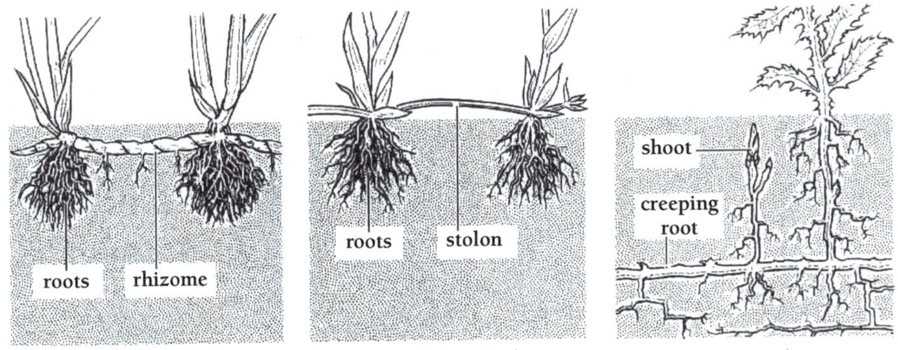

FIGURE 1-15.

Many weeds reproduce by means of one or more types of underground and aboveground rooting structures, including those illustrated here.

plant. Tubers also grow underground. They are enlarged fleshy growths that arise from stems or roots. A stolon is a stem that grows horizontally above the ground surface and can develop roots at its nodes.

Roots. Roots absorb water and nutrients from the soil and store food. The creeping roots of some plants give rise to stems. Some weeds (dandelions, for example) have thick, elongated taproots from which short lateral rootlets grow. Other weeds such as grasses have a network of fine branching fibrous roots. Roots can provide helpful clues in weed identification.

Fruits and Seeds. A fruit is the ripened ovary of a plant's flower. Seeds are the primary means of reproduction for annual and some perennial plants. Fruits and seeds are useful structures when you are identifying weeds in that they are unique in their shape, size, markings, and color.

INVERTEBRATES

Invertebrates are animals without backbones (vertebrae). These include the arthropods (insects, spiders, mites, and their relatives), nematodes and all other worms, and snails and slugs. Pest invertebrates affect turf and landscape plants in many different ways. Typically, invertebrate pests in a landscape feed on growing plants. Some transmit disease organisms to the plants.

Arthropods

Arthropods make up one of the largest groups in the animal kingdom. The word *arthropod* means "jointed foot" and refers to animals with an external skeleton and jointed body parts. Insects, spiders, ticks, and mites all belong to this group, as do crabs, crayfish, shrimp, and lobsters. Centipedes, millipedes, scorpions, and sowbugs are arthropods, too. Only about 3% of the arthropod species that occur in the United States can be classified as pests.

Sometimes insects, mites, and spiders may be pests in one situation but beneficials in another. For example, spiders are pests when they disturb or endanger people, but they can also help to control certain plant-feeding insects and other arthropod pests.

How to Identify Arthropods

Arthropods range in size from several inches long right down to microscopic. You can identify most important arthropod pests by comparing them to photographs or drawings such as those in *Pests of Landscape Trees and Shrubs: An Integrated Pest Management Guide, Second Edition* or the UC Integrated Pest Management Program's Web site

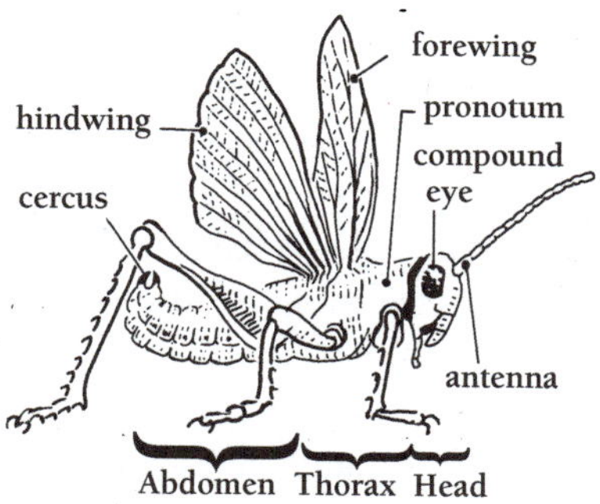

FIGURE 1-16.

The body parts of this grasshopper represent the general structures that you usually see on most adult insects. Some adult insects are wingless, however.

(http://www.ipm.ucdavis.edu/). Use a microscope or hand lens to examine small specimens. If you plan to send pests to an identification specialist, see Sidebar 3 for instructions on how to prepare them for shipment.

People identify pest insects and mites by distinguishing between the various types of body structures that are characteristic of different groups. Some insects undergo changes in their body form during their lifetime (for example, caterpillars turn into moths or butterflies). Adults of many pest insects have wings that are useful for identification. Insect mouthpart types are also helpful for identification: usually insects have mouthparts that are specialized for sucking, biting, or chewing, or a combination of these.

Body Structure

Insects. The body of an adult insect has three distinct parts: the head, the thorax, and the abdomen (Figure 1-16). Eyes and one pair of antennae arise from the head. Usually the head has several pairs of appendages that make up the mouthparts. The thorax has three pairs of legs and often one or two pairs of wings. Insects have segmented abdomens. Sometimes the wings partially or entirely cover the abdomen when folded. Some insects have appendages that extend from the tip of the abdomen, such as pinchers, a stinger, or other structures.

Spiders. Spiders have two major body parts (Figure 1-17). The head and thorax combine into one section called

SIDEBAR 3

Sampling and Sending Arthropods for Identification

SAMPLING

- ☐ Collect plant-feeding insects and mites by snipping off portions of foliage or stems containing the pest and placing these into a plastic bag.
- ☐ Use an insect net to collect flying insects.
- ☐ For other insects, shake plant foliage onto a light-colored cloth sheet and funnel arthropods into a plastic bag or into a glass or plastic jar.
- ☐ Keep all collected specimens cool by placing them into an ice chest or refrigerator.

PREPARATION

- ☐ Place insects, mites, and other arthropods into glass vials with 70% isopropyl alcohol ("rubbing alcohol" also may be used).
- ☐ Vials must be sealed so that alcohol cannot leak out.
- ☐ Include more than one individual of each species whenever possible. If other life stages are present (eggs, larvae, pupae, adults), include representative samples of these.
- ☐ If the pest is associated with plants, send samples that show pest damage. Do not put plant material in the alcohol. Keep plants as fresh as possible by keeping them cool.

LABELING

Attach a label to the outside of each vial or plant sample bag. Include the following information on labels:

- ☐ Your name, address, and telephone number.
- ☐ Date specimens were collected.
- ☐ Name and variety of host plant, if applicable.
- ☐ Whether pest is in commercial, residential, nursery, or other setting. If the pest is not associated with plants, describe the site where it was collected.
- ☐ Location of area where specimens were collected, including names of nearest crossroads.

SHIPPING

- ☐ Contact the person or laboratory that will receive samples to determine the best method of shipping and to inform them that samples will be arriving.
- ☐ Pack samples in a sturdy, well-insulated container to prevent crushing or heat damage.
- ☐ Mark the package clearly and request that the shipper keep it under cool conditions.
- ☐ If plant material is included, ship packages early in the week so they will arrive before a weekend.

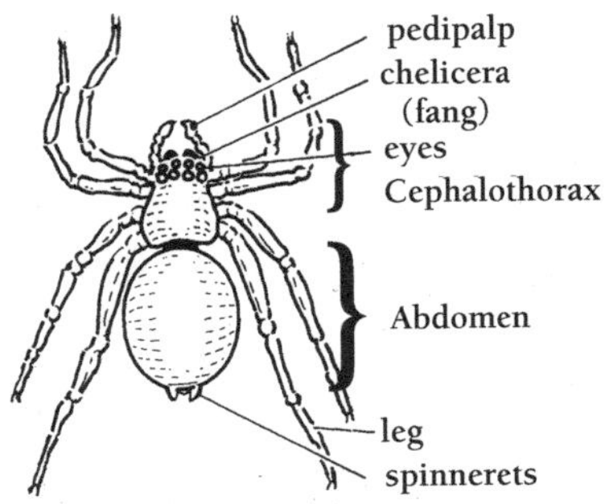

FIGURE 1-17.

Spiders have two main body regions rather than the three found in insects. They have four pairs of legs, a pair of pedipalps, and a pair of chelicerae (fangs). At the end of the abdomen is a cluster of spinnerets, part of the spider's web-producing mechanism.

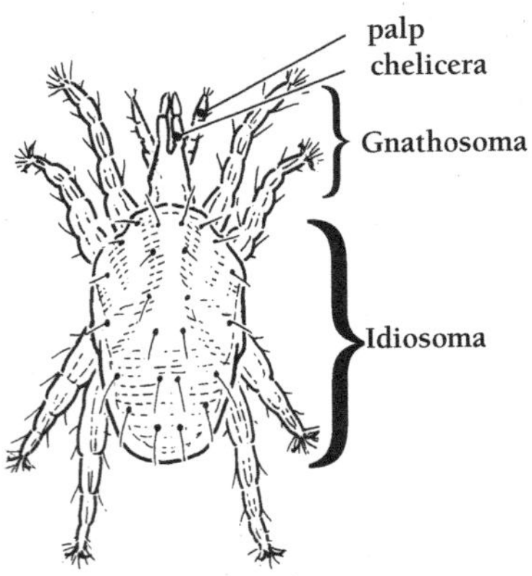

FIGURE 1-18.

The body arrangement of mites and ticks is different from that of an insect or spider. Adults mites and ticks generally have four pairs of legs; immature forms usually have three pairs.

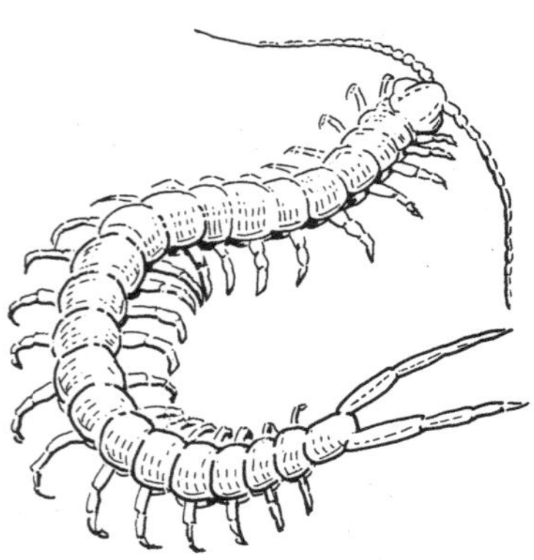

FIGURE 1-19.

Centipedes have many body segments, most of which give rise to pairs of legs. Centipedes have a pair of poisonous fangs that arise from the head segment.

a cephalothorax. The cephalothorax gives rise to four pairs of legs. In addition, eyes and mouthparts, including a pair of chelicerae (fangs), arise from the cephalothorax. The remaining body part is a non-segmented abdomen that terminates with several pairs of spinnerets, the spider's web-spinning organs.

Mites. A mite has only two body segments, but they are different from those of a spider (Figure 1-18). A small head (called a gnathosoma) with a few mouth appendages is attached to a large, combined thorax and abdomen called an idiosoma. Adult mites have four pairs of legs that arise from the idiosoma, and immature forms usually have only three pairs. Plant pest mites feed by sucking fluid from plant cells.

Other Arthropods. Other arthropod groups have varying arrangements of body parts and legs. For example, centipedes and millipedes have many body segments and large numbers of legs (Figure 1-19). Scorpions have a long, segmented tail that terminates in a sting apparatus.

Developmental Stages

Arthropods hatch from eggs and increase in size by shedding their outer body covering, after which they grow a new, larger one. This is the process of molting. The period between one molt and the next is an instar. Immature arthropods pass through several instars before becoming adults.

The developmental changes from egg to adult that take place in most insects are unique among the arthropods. In many insect orders, the young look entirely different from the adults. Immature insects in these orders undergo a change in form known as complete metamorphosis. These insects pass through a larval (usually wormlike) stage and a pupal (resting) stage before they become adults (Figure 1-20A).

Examples include moths and butterflies, beetles, and ants, bees, and wasps. In other insect orders the young resemble adults, but lack the adults' wings. Examples include the true bugs, grasshoppers and crickets, and cockroaches. As these insects grow, they pass through a gradual metamorphosis. If the insects have wings as adults, the young insects develop external wing buds and these gradually grow until the insects reach their adult stage (Figure 1-20B).

SPIDERS: Class Arachnida—Order Araneae

Important Characteristics. A spider has four pairs of legs and an unsegmented abdomen. The abdomen attaches to the cephalothorax, the other main body part, by a narrow waist. At the tip of the abdomen are spinnerets, special organs used to produce different types of webbing.

Life Cycle. Immature spiders hatch from eggs and pass through several instars before becoming adults. Each instar begins by molting as the spider grows. Most spiders live for two or three years, although some species have lifespans as long as 20 to 30 years. Females generally live longer than males. Spiders eat insects and other small arthropods.

Where You Find Them. Spiders are found in buildings and in most types of landscape environment. They frequently live among plants.

Damage. In California, black widow spiders and a few large spiders can inflict venomous bites. Sometimes these bites may require medical attention. Spiders leave unattractive webbing that gathers dust and debris on plants.

Beneficial Aspects. Spiders are general predators of insects and other arthropods and they contribute to the natural control of many pest species. For example, orb-weaving spiders snare and feed on flies and moths.

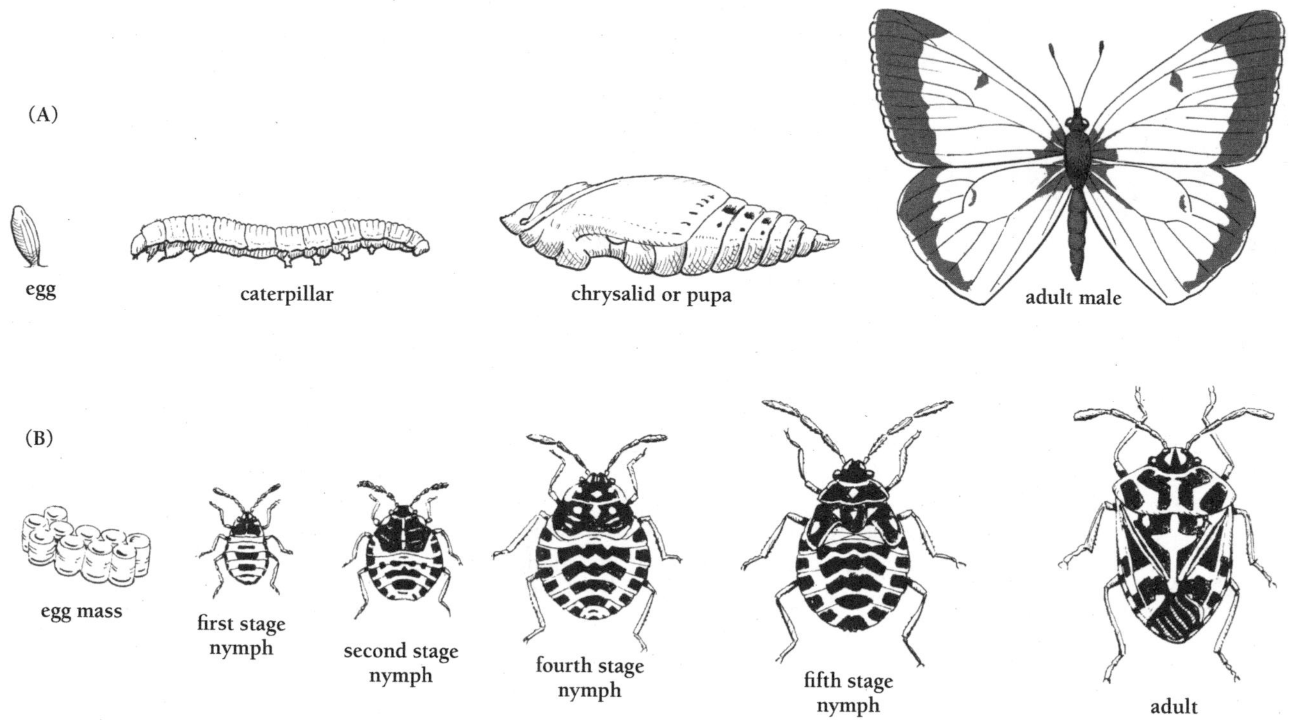

FIGURE 1-20.

Some orders of insects undergo complete metamorphosis (A). Other orders undergo incomplete metamorphosis (B).

MITES: Class Arachnida—Order Acari

Important Characteristics. The combined abdomen and thorax (idiosoma) of a mite is broadly joined to the head (see Figure 1-18). Adults usually have four pairs of legs, while immature mites most often have three pairs or fewer. Some mite species produce fine webbing from silk glands located near the mouth. Most mites are very small and are difficult to see without the aid of a hand lens or microscope.

Life Cycle. Mites hatch from eggs and pass through several instars before they become adults. Immature mites resemble the adults. Mites usually develop from egg to adult in a matter of days to a few weeks. Some species overwinter as adults; others overwinter as eggs.

Where You Find Them. Plant-feeding mites live on (mostly) lower or upper leaf surfaces and on new growth.

Damage. Mites feed on plants by puncturing plant cells with their mouthparts and sucking the fluids. Plant-feeding mites often produce serious economic or visible damage, including leaf stippling, bleaching, defoliation, and deformation. Gall mites cause blistering on leaves and gall formation on twigs. Some plant-feeding mites transmit disease-causing microorganisms (pathogens). See *Pests of Landscape Trees and Shrubs, Second Edition,* for examples of plants that can be attacked by various mites.

Beneficial Aspects. Several mite species prey on other pest mites and are not pests themselves. These sometimes play an important role in natural and biological control programs.

ORTHOPTERANS: Class Insecta—Order Orthoptera (crickets, grasshoppers, locusts, katydids)

Important Characteristics. The bodies of immature orthopterans resemble those of adults, except that where

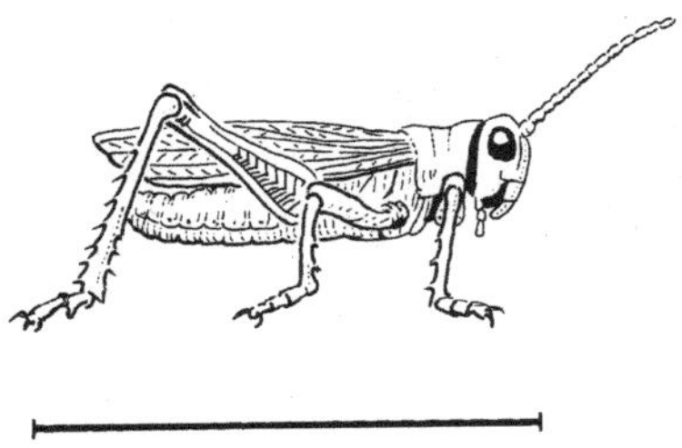

DIFFERENTIAL GRASSHOPPER

FIGURE 1-21.

Grasshopper—Order Orthoptera.

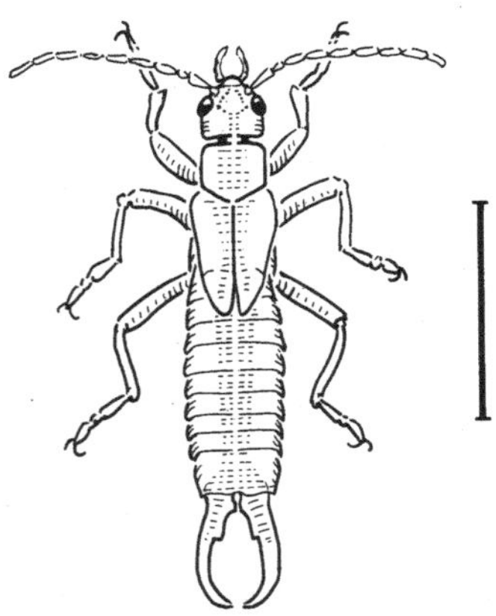

EUROPEAN EARWIG

FIGURE 1-22.

Earwig—Order Dermaptera.

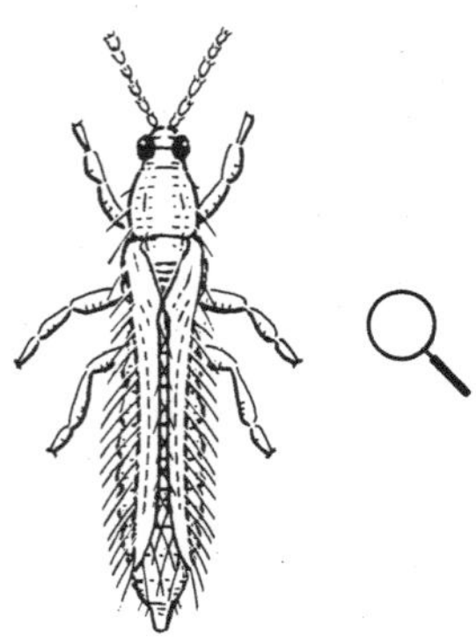

WESTERN FLOWER THRIPS

FIGURE 1-23.

Thrips—Order Thysanoptera.

adults of many species have wings, the immatures do not. Orthopterans have chewing mouthparts and many species' adults have wings, although most of those are not good flyers (Figure 1-21). Some species are wingless. Most insects in this order have powerful hind legs that they use for jumping. Crickets, locusts, and katydids make distinguishing chirping or buzzing sounds. Many grasshopper species make clicking or buzzing sounds when they fly.

Life Cycle. Orthopterans hatch from eggs that adult females glue to plant surfaces, insert into plant tissues, lay in soil, or lay freely on the ground. Immature forms do not have wings. They pass through a series of instars before reaching the adult stage. Most orthopterans live one or more years.

Where You Find Them. Orthopterans occur in most landscaped areas. Many live in the soil, while others live in trees or shrubs.

Damage. Grasshoppers, locusts, crickets, and katydids feed on plants. They sometimes cause serious injury or defoliation in fruit crops, rangeland, and vegetable gardens, but their chewing usually does not seriously injure landscape or turf plants in California.

EARWIGS: Class Insecta—Order Dermaptera

Important Characteristics. Earwigs are elongated, sometimes wingless insects with characteristic pinchers (cerci) that extend from the end of the abdomen (Figure 1-22). They use these pinchers for defense and to catch prey. Female earwigs generally have straight cerci, while males' cerci are curved. Earwigs have chewing mouthparts.

Life Cycle. Earwigs hatch from eggs deposited in a nest in the soil. Adult females protect their eggs and their newly hatched young. Earwigs pass through gradual metamorphosis, developing from eggs to adults in two to three months. They hibernate as adults.

Where You Find Them. Earwigs usually nest in the ground and hide under boards, stones, and ground litter. Earwigs are very common insects in some parts of California.

Damage. In landscaped areas, European earwigs feed on flowers. They also feed on vegetables, ripe fruit, and garbage. Some earwig species are predaceous on other insects.

THRIPS: Class Insecta—Order Thysanoptera

Important Characteristics. Thrips are tiny, elongated insects with two pairs of wings (Figure 1-23). Their wings have a fringelike appearance. Thrips have modified sucking-rasping mouthparts.

Life Cycle. Thrips hatch from eggs and undergo incomplete metamorphosis. Most species pass through four instars. Thrips actively feed during their first two instars. The final instars are more of a resting stage, and often take place in the soil. Wings are present after the thrips' final molt.

Where You Find Them. Thrips commonly infest plants, often in flowers and on tender, developing parts of leaves and fruits.

Damage. Thrips puncture plant cells and then suck up the fluid that escapes. This type of feeding damages cells, resulting in stunted growth, leaf discoloration, and deformed fruit, leaves, flowers, and shoots. Some thrips are serious pests in greenhouses, gardens, interiorscapes, and landscaped areas, and they may transmit plant disease organisms. Some of the plants that are susceptible to thrips damage are azaleas, *Hypericum* (gold flower, St. Johnswort), English and Grecian laurel, photinia, rhododendron, toyon, impatiens, petunias, roses, fruit trees, pomegranate, and fig.

Beneficial Aspects. A few thrips species are predatory. Predatory thrips play an important role in the natural control of several plant pests, including aphids and mites.

TRUE BUGS: Class Insecta—Order Heteroptera (bed bugs, plant bugs, damsel bugs, assassin bugs)

Important Characteristics. You can recognize most true bugs (Figure 1-24) by the triangular shape of the scutellum when seen from above. A true bug also has a long, needlelike beak that folds under the body (piercing-sucking mouthparts). True bugs vary in size depending on the species: some are nearly two inches long. They have two pairs of wings and most fly well. The second pair of wings is not visible while the insect is at rest. They sometimes have brightly colored wings.

Life Cycle. True bugs undergo gradual metamorphosis after they hatch from eggs. The young resemble adults, but without wings. The number of instars and amount of time to reach maturity vary among the many true bug species.

Where You Find Them. True bugs feed on plants and animals, depending on the species. Most species wander about, searching for appropriate hosts for food. There are several aquatic species. Plant-feeding species such as lace bugs suck sap from the underside of plant leaves.

Damage. Plant-feeding true bugs damage plant cells. This causes deformed flowers, fruits, leaves, and stems. Some bugs also inject chemicals into the host plant that prevent or alter the plant's normal growth. Bugs are often serious plant pests. Trees that may be attacked by plant-feeding true bugs include alder, ash, birch, box elder, fruit and nut trees, maple, poplar, sycamore, and willow. Chinch bugs may attack turfgrass and most herbaceous plants. Plants that other bugs may attack include azalea, ceanothus, cotoneaster, coyote brush, photinia, pyracantha, and toyon.

Beneficial Aspects. Some species are predators of other insects, including many insect pests. Examples of these are assassin bugs, bigeyed bugs, and minute pirate bugs.

HOMOPTERANS: Class Insecta—Order Homoptera (aphids, leafhoppers, cicadas, psyllids, whiteflies, mealybugs, scales, phylloxeras, spittlebugs, treehoppers)

Important Characteristics. Homopterans are a diverse group of somewhat soft-bodied insects. Most are winged, although some, such as aphids, have both winged and non-winged forms. They all have piercing-sucking mouthparts. Figure 1-25 shows some of the different homopterans.

Life Cycle. Homopterans undergo incomplete metamorphosis. Because this group is so diverse, it is not possible to describe a typical life cycle. Some, such

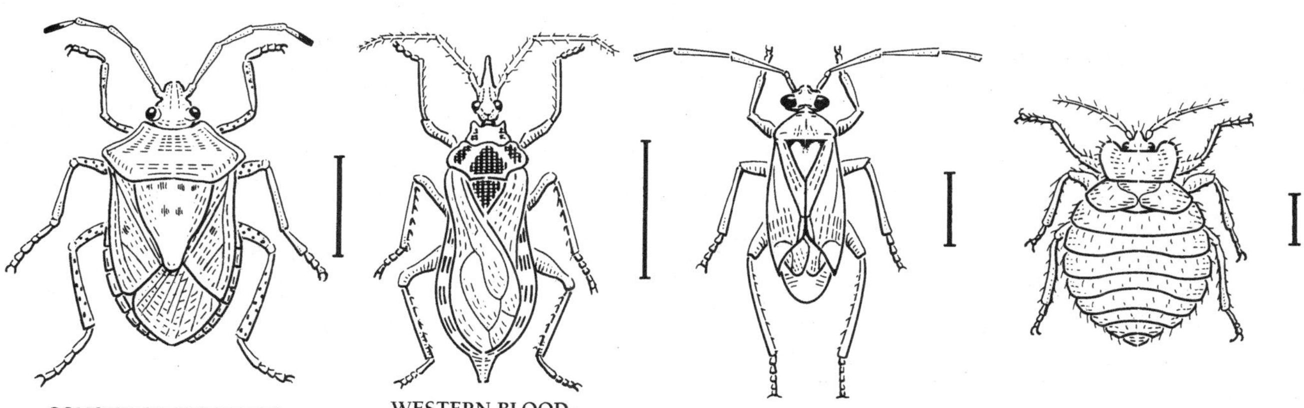

FIGURE 1- 24.

True bug—Order Heteroptera.

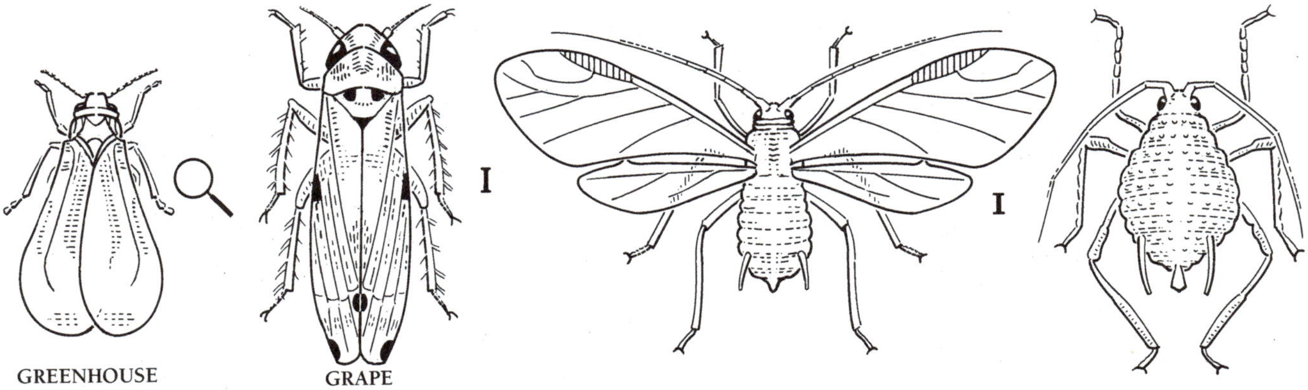

FIGURE 1-25.
Whitefly, leafhopper, and aphid—Order Homoptera.

as the periodical cicada, have life cycles that span as far as 17 years, but many homopterans live less than one year.

Where You Find Them. Homopterans are plant feeders, so they are usually found on or near plants. Some occur in greenhouses and on indoor plants.

Damage. Homopterans pierce plant tissues and suck out liquids. Feeding usually causes deformity in leaves and fruit, loss of plant vigor, stunted growth, and dieback of plant parts. Many homopterans excrete a sticky substance called honeydew that supports the growth of black sooty mold fungi. The honeydew and sooty mold create a sticky and unsightly mess on trees, sidewalks, vehicles, and other surfaces beneath the plant. Some aphid species transmit disease-causing pathogens to host plants. See *Pests of Landscape Trees and Shrubs, Second Edition* for examples of plants attacked by various aphid species.

FIGURE 1-26.
Beetle—Order Coleoptera.

COLEOPTERANS: Class Insecta—Order Coleoptera (beetles, weevils, wood borers)

Important Characteristics. Adult beetles, weevils, and wood borers range from "pinhead" size to several inches in length, depending on the species (Figure 1-26). Most adults have elytra, a pair of hardened front wings. These completely cover the top of the abdomen and hide the insect's second pair of wings when it is not flying. Larvae are very diverse in appearance. Both larvae and adults have chewing mouthparts.

Life Cycle. Beetles, weevils, and wood borers undergo complete metamorphosis. Larvae pass through several instars before pupating, and pupation usually takes place in the soil. Some adults survive adverse weather by going into a resting or dormant phase—hibernation during cold winter periods or aestivation during hot or dry periods.

Where You Find Them. Several types of beetle, weevil, and wood borer infest plants and may feed on leaves, stems, or bark as adults or larvae.

Damage. The feeding activities of pest beetles, weevils, and wood borers damage, disfigure, or destroy trees and other plants. They chew on leaves, flowers, twigs, stems, and roots. Some bore through the bark of trees and shrubs. See *Pests of Landscape Trees and Shrubs, Second Edition* for examples of plants that are attacked by various coleopterans.

Beneficial Aspects. Many beetle species prey on other insects. These beetles contribute to the control of plant-feeding pests such as aphids and mites. Common predatory beetles that feed on plant pests include the convergent lady beetle, the twicestabbed lady beetle, and the vedalia beetle.

LEPIDOPTERANS: Class Insecta—Order Lepidoptera (butterflies, moths, skippers)

Important Characteristics. Adult butterflies, moths, and skippers (Figure 1-27) are different from other insects in that they have large, scale-covered, often brightly colored wings. Their larvae are caterpillars with chewing mouthparts. Adults have modified mouthparts in the form of a coiled tube that they extend to suck up liquids. You can distinguish butterflies and skippers from moths by looking at their antennae. Moths have tapering hairlike or feathery antennae, while butterflies and skippers usually have clubbed antennae (Figure 1-28). Skippers have shorter, thicker bodies and smaller wings that set them apart from butterflies. Moths are primarily active at night, and butterflies and skippers fly mostly during the day.

Life Cycle. Moths and butterflies undergo complete metamorphosis. After hatching from eggs, larvae pass through several instars and then enter the pupal stage and change into winged adults. Pupation takes place in cocoons that the larvae build and attach to plant foliage or other objects. Pupation sometimes takes place in the soil. Many pest lepidopterans overwinter as pupae. Several pest species produce three or four generations per year. These insects do their damage in the larval stage, when they feed on plant leaves or flowers.

Where You Find Them. Lepidopteran pests occur on or in plant parts. Adult moths are commonly attracted to lights at night.

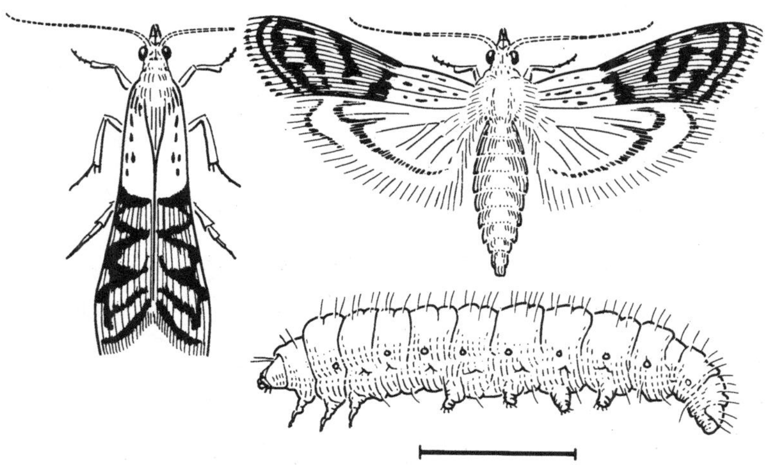

FIGURE 1-27.
Moths—Order Lepidoptera.

Damage. Moth larvae can be serious plant pests. Their feeding can leave holes in leaves or flowers or can destroy entire leaves, depending on the pest species. Clearwing moths attack trunks and branches of trees. The larvae may leave excrement or frass on plant surfaces and this may fall to the ground beneath infested plants. See *Pests of Landscape Trees and Shrubs, Second Edition* for information about moth species that bore into plants and moth species that feed on leaves, as well as common host plants for both.

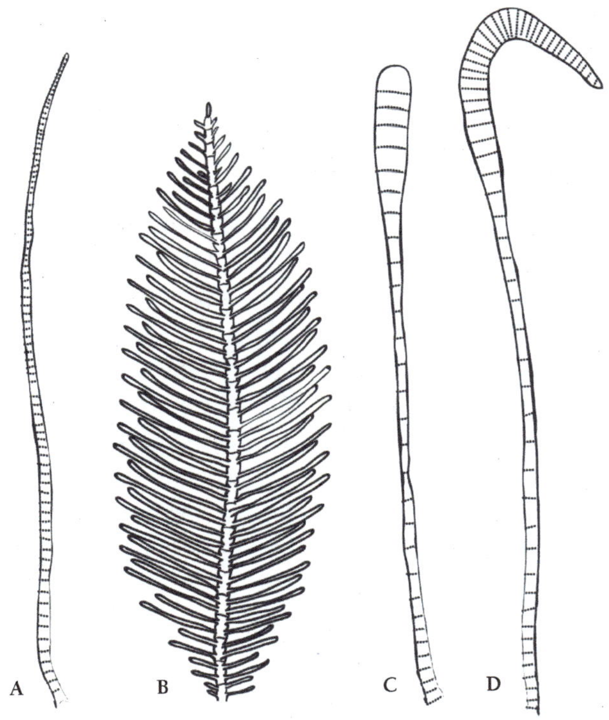

FIGURE 1-28.

Moths have tapering hairlike antennae (A) or feathery antennae (B), while butterflies (C) and skippers (D) usually have clubbed antennae.

Beneficial Aspects. People consider most butterflies desirable in the landscape, but to have butterflies one must also be willing to tolerate their plant-chewing larvae.

HYMENOPTERANS: Class Insecta—Order Hymenoptera (bees, wasps, ants, sawflies, horntails)

Important Characteristics. One way to distinguish hymenopterans from insects of other orders is to examine the point where the abdomen joins the thorax. Most hymenopteran adults have a very narrow "waist" (Figure 1-29). Sawflies (which are broad-waisted, stingless wasps) are exceptions to this rule.

In their winged form, adult hymenopterans have two pairs of transparent wings that are sometimes colored. Larvae are grublike. Larvae

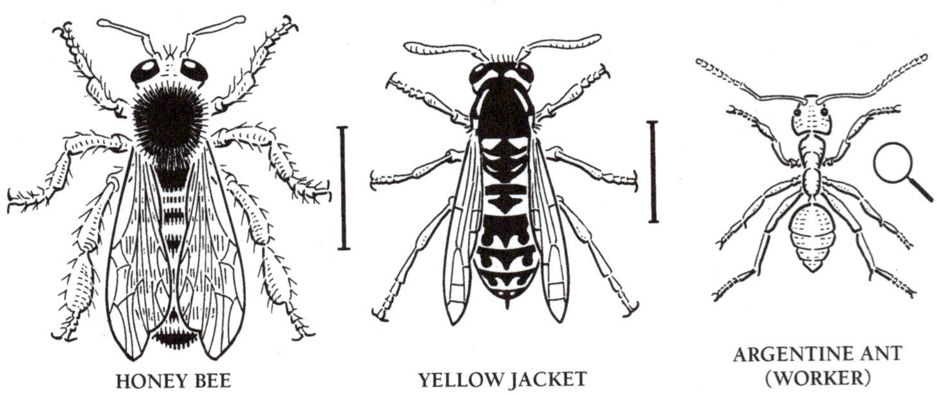

FIGURE 1-29.

Bee (left), ant (right), and yellow jacket (center)—Order Hymenoptera.

and some adult hymenopterans have chewing mouthparts. Certain bee, wasp, and ant species are venomous and are equipped with a stinger at the tip of the abdomen. Several species are social rather than solitary. In species that have a social structure, large numbers of individuals generally live together in a colony and members of the colony perform different tasks depending on their caste.

Life Cycle. Hymenopterans undergo complete metamorphosis. The larvae of some wasps may be internal or external parasites of insects, spiders, and other arthropods, and some species' larvae are plant parasites. In colonies of social hymenopteran species, adults forage for food for their larvae. Colonies may live for many years, although their individuals survive for only a few months to one or two years. Some species produce several generations per year.

Where You Find Them. Adult hymenopterans forage for food on flowers and other plant parts, and are attracted to sweets and meat. Social species live in nests in the ground or in buildings, trees, and other structures.

Damage. Many species of Hymenoptera are venomous and inflict painful stings that, for some people, may be fatal. Certain sawfly species are serious plant pests. Sawflies mostly infest conifers but sometimes infest other plants such as fruit trees and roses. See *Pests of Landscape Trees and Shrubs, Second Edition* for information about the plant hosts of several sawfly species. Many bees and wasps are nuisances around public areas. Ants are persistent pests around plants and some bees and ants damage wood structures. Some species, such as Argentine ants, harvest the honeydew produced by aphids or scale insects and use this as food for their larvae. These ants protect the aphid or scale insect colonies from attack by parasitic wasps.

Beneficial Aspects. Bees and other species of Hymenoptera are important pollinators of agricultural and horticultural plants. Parasitic wasps play an extremely important role in the control of pest insects. Some are egg parasites and others attack the larval or adult stages of their hosts. Adults of some species are predaceous.

Nematodes

The nematodes are a large group of species of unsegmented worms. At most stages of development, nematodes are pointed at both ends and quite distinct in appearance from segmented worms such as earthworms. Mature females of some species have various round-ended shapes. Most species are very small and can only be seen with a microscope. While some nematodes are plant parasites and others are animal parasites, most species are not parasites at all. Certain beneficial nematodes attack insect pests.

Most plant-parasitic nematodes are internal or external parasites of plant roots, although some feed on foliage and other aboveground parts. Some root-feeding species inject the plant tissues with salivary secretions that cause root cells to enlarge and form knots or galls in which the nematodes live and feed (Figure 1-30). Other species feed on root tips or in inconspicuous lesions on the root surface. Root damage from nematodes prevents the plant from getting the water and nutrients required for normal growth. Plant symptoms include stunting, loss of vigor, yellowing, and general decline. The damage to foliage may result in stunting or the production of galls in place of seeds. Some nematodes transmit viruses to plants. Although nematodes seldom actually kill plants, nematode-infested plants are highly susceptible to injury from other pests or environmental stresses.

Nematodes are difficult to identify because of the large number of species and because most are microscopic.

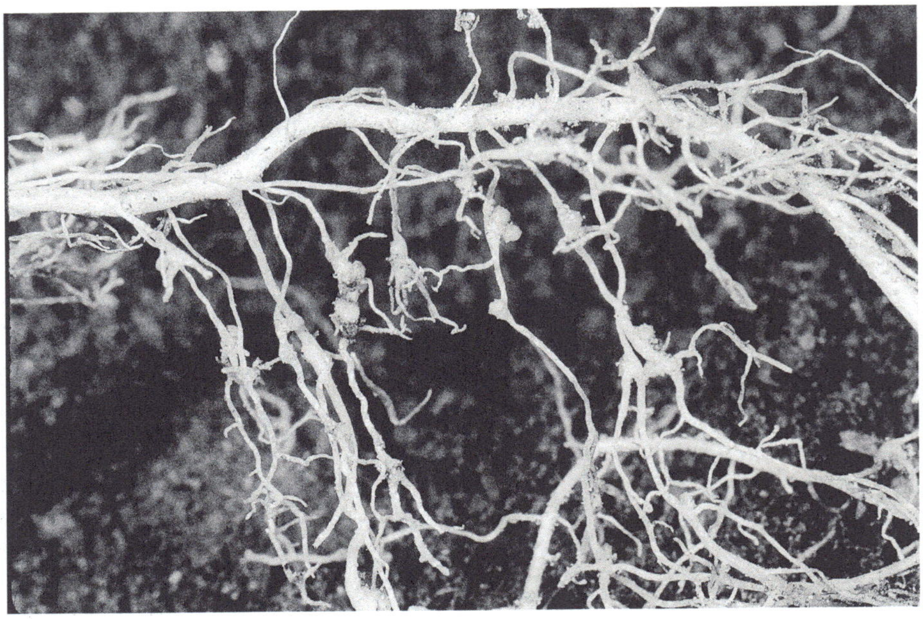

FIGURE 1-30.

The rounded nodules on the roots of this plant are galls caused by nematodes. Gall-forming nematodes inject plant tissues with salivary secretions that cause root cells to enlarge and form these knots or galls in which the nematodes then live and feed.

To identify plant pest nematodes, collect samples of soil, roots, and other affected plant parts. Besides taking samples from infected areas of a planting, take separate samples from areas that appear normal. Place root samples in the same bag as the soil to prevent them from drying, but keep aboveground plant parts separate from roots and soil. Send your samples to a laboratory for identification (see Sidebar 4 for information on sampling and sending samples for identification).

After you work in an area infested with nematodes, always clean your shoes and scrub your tools and equipment thoroughly with hot water or a chlorine solution. Be sure to remove all soil from the tools. This helps reduce the spread of pests to other areas.

Damage. Most of the important nematode plant pests damage roots or other underground parts such as tubers or rhizomes. A few, however, feed aboveground and damage stems, leaves, flowers, or seeds, and can be significant plant pests. Damage to the root system prevents an infected plant from obtaining water and nutrients. Infected plants wilt under temperature and water stress, may become stunted or yellow, and are susceptible to infections by secondary pathogens such as fungi.

Snails and Slugs: Phylum Mollusca

Snails and slugs (Figure 1-31) belong to the phylum Mollusca, along with clams, scallops, abalone, octopus, and squid. Snails have shells, distinguishing them from slugs, which have no visible shells. In general you need not identify a particular snail or a slug species in order to achieve successful control. You should, however, confirm that these pests are the actual cause of observed damage before you start to apply management measures. Endangered species laws protect some native snail and slug species, so these cannot be destroyed.

Important Characteristics. Snails prefer cool, moist surroundings. With their shells to protect them from heat and dryness, however, they are able to live in many different environments.

SIDEBAR 4

Sampling and Sending Plants for Identification of Plant-Infesting Nematodes

SAMPLING

- ☐ Take random samples, but sample in a consistent manner. Mark off the sampling area into grids. Randomly select the gridded areas from which you will collect samples.
- ☐ Keep samples collected from areas where you suspect disease separate from samples taken in healthy areas.
- ☐ Dig up and include roots and soil of diseased and healthy plants. Take root samples from below the level of surface feeder roots, since temperature and moisture fluctuations at the surface will affect the nematode species living there.
- ☐ Include partially rotted or decayed roots.
- ☐ Include aboveground plant parts if you suspect that the nematodes are infesting these areas. Keep aboveground parts separate from root and soil samples.

PREPARATION

- ☐ Place samples in clean plastic bags and keep them out of the sunlight. Keep samples in an insulated ice chest or refrigerator until they can be shipped. Do not freeze.
- ☐ Pack samples in a well-insulated carton or a disposable foam ice chest. Make sure that the container is sturdy enough to prevent damage to its contents.

LABELING

Attach a label to the outside of each sample bag. Include the following information on labels:

- ☐ Your name, address, and telephone number.
- ☐ Date samples were collected.
- ☐ Plant types, including variety.
- ☐ Portion of planted area that the sample represents. Include a map if necessary.
- ☐ Location of property where samples were taken (names of nearby crossroads).
- ☐ Your observations on previous problems and on when the present problem was first detected.

SHIPPING

- ☐ Contact the person or laboratory that will receive samples to determine the best method of shipping and to inform them that samples will be arriving.
- ☐ Pack samples in a sturdy, well-insulated container to prevent crushing or heat damage.
- ☐ Mark the package clearly and request that the shipper keep it under cool conditions.
- ☐ Ship packages early in the week so they will arrive before a weekend.

A snail can seal itself into its shell and become dormant for up to four years during dry periods. Slugs, which have only tiny, rudimentary shells, are vulnerable to high temperatures and dry weather, so they prefer cool, damp locations.

Life Cycle. A snail or slug will lay between 10 and 200 eggs in a mass beneath the surface of the soil. They lay egg masses several times each year from spring through fall. In cooler areas, egg-laying activity stops during the winter. It takes from several months to three years for newly hatched snails to reach maturity. Slugs may reach maturity in three to six months.

Where You Find Them. Snails and slugs live in damp areas, soil litter, and plant foliage. They are usually nocturnally active. During the day they hide under boards or stones, and among ivy, dense shrubbery, or damp refuse. At night and early in the morning or during cool, damp periods, they forage for food. They often return to the same resting area each day unless the area becomes too dry or has been disturbed.

Damage. Snails and slugs are not serious pests in landscape settings, although they feed on many plant species and they favor tender, young foliage and flowers. They are also pests in greenhouses. Snails and slugs seek out

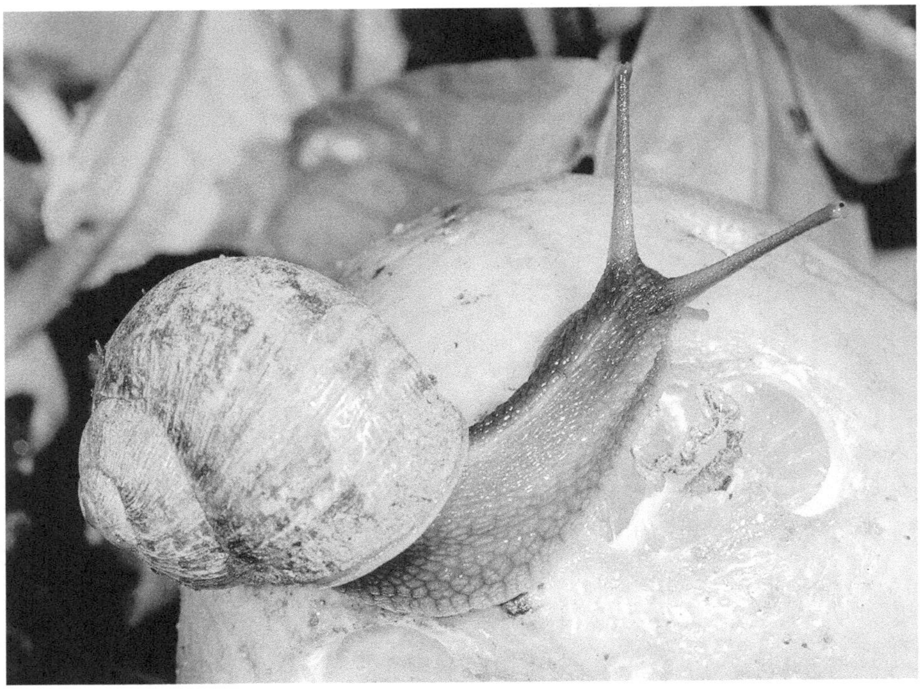

FIGURE 1-31.
Brown garden snail—Phylum Mollusca.

succulent foliage near the ground and are primarily pests of seedlings, herbaceous plants, and other low-growing vegetation. Besides their feeding damage, both snails and slugs leave slime trails that detract from the appearance of foliage.

Beneficial Aspects. Some species of snails are predaceous and can be useful for biological control of pest snails. For example, decollate snails feed mostly on other snails and decaying organic matter. Decollates also eat young seedlings, however, so it may not be a good idea to release them in some landscape areas. In many citrus groves, the establishment and proper maintenance of decollate snail populations has permanently reduced brown garden snail populations to insignificant levels over the course of 4 to 10 years. Decollate snails introduction is permitted only in certain San Joaquin Valley and Southern California locations. The introduction of decollate snails is illegal in other areas because they may affect native snail and slug populations.

VERTEBRATES

Vertebrates (phylum Chordata) are animals with internal skeletons and backbones. They include fish, amphibians (frogs, toads, and salamanders), reptiles (turtles, lizards, and snakes), birds, and mammals. They become pests if they

- spread disease-causing pathogens (such as plague or rabies)
- damage plants
- make burrows or mounds that become safety hazards
- interfere with the activities or needs of people

Laws and regulations on endangered species, migrant species, and other wildlife protect many vertebrate species, so it is important that you positively identity the pest species before you begin any management program. Probably fewer than a dozen vertebrate species constitute a pest concern in landscapes.

Identify vertebrates by comparing their damage symptoms, nests, or droppings to reference photographs and drawings. For help in identifying vertebrates, contact the U.S. Fish and Wildlife Service or the California Department of Fish and Game. These agencies require permits for control of some vertebrate species. In addition, the California Department of Fish and Game may require a trapper's license for anyone trapping certain furbearing mammals for hire.

Birds

People usually appreciate birds because of their song, beauty, or predaceous habits. Many birds catch insects and some kill snakes, rats, mice, and other pests. Birds themselves often become pests, however, when they leave unsightly droppings on plant surfaces and other objects in their roosting areas, especially when they roost in massive numbers in trees or on buildings. In addition, some species eat or damage parts of desirable plants.

Mammals

Certain mammals are seen as pests when they interfere with the activities of people or cause harm to desirable plants. In addition, rodents such as rats (Figure 1-32), mice, and squirrels often carry fleas, mites, lice, or ticks that harbor and transmit disease organisms. Squirrels in parts of California, for example, may carry fleas that transmit the plague bacterium, *Yersinia pestis.* Infected skunks and bats and other infected mammals can transmit the rabies virus. Larger mammals such as foxes, coyotes, raccoons, mountain lions, and bears are occasional pests when they endanger people or prey on pets. Foraging deer can be serious landscape pests. *Wildlife Pest Control around Gardens and Homes* (UC ANR Publication 21385) is a helpful guide that can help you identify pest mammals and decide how to manage them.

DISEASE AGENTS

Another important group of pests includes certain plant disease agents. By altering or interfering with chemical processes that take place within the plant cells, these agents produce disease symptoms or disorders.

Biotic disease factors are pathogens

FIGURE 1-32.

Rats can be pests when they damage trunks, limbs, and foliage. They may carry certain fleas, mites, lice, or ticks that harbor and transmit disease organisms. This is a Norway rat, Rattus norvegicus.

that are spread from one host to another and produce disease symptoms. The ability of the pathogen to spread and infect depends on climatic factors, host susceptibility, and the pathogen's potency. The host's genetic makeup and nutritional state also influence this ability. Several different types of microorganism produce diseases in plants. The most important in landscapes are bacteria, fungi, and viruses.

Identifying Disease-Causing Pathogens

Many pathogens are submicroscopic, so small that identification is very difficult. In some cases, an electron microscope or complex biochemical tests are the only means available to confirm the identity of the pathogens. It may even be necessary to grow the organism in the laboratory or to inject plants with the suspected pathogen in order to make a positive identification. This sort of testing requires trained technicians and specialized laboratory equipment. Sidebar 5 gives information on how to use laboratory identification services and how to send material for examination.

Learning to recognize plant disease symptoms will help you sort through and reject some pathogens as possible causes of the observed symptoms. You need to know the species of the infected host plant before you can correctly identify the pathogen causing the damage. Use host tables such as those in *Pests of Landscape Trees and Shrubs, Second Edition* to narrow down the causes of common problems on your plants. In some cases, specific disease-causing pathogens are more prevalent in a particular location on particular plant species or genera, or under other specific conditions. Knowledge of these limitations will also help you to narrow the list of possible pathogens.

Fungi produce distinctive structures such as molds or spore-producing bodies (mushrooms, for instance) that can help you identify the organism. Lesions or rotted areas of infected plants may provide clues to the type of organism causing the damage. Inspect the plant's

SIDEBAR 5

Sampling and Sending Plants for Identification of Disease-Causing Pathogens

SAMPLING

☐ Select plants that are most representative of the observed disease symptoms. Collect several plants.

☐ Include roots. To do it, dig up plants and shake off the soil.

☐ Place plants in plastic or paper bags. Keep plant material cool. Put them in an ice chest while you are in the field. Store samples in a refrigerator until you can ship them.

LABELING

Attach a label to the outside of each sample bag. Include the following information on labels:

☐ Your name, address, and telephone number.

☐ Date samples were taken.

☐ Crop or plant type, including variety.

☐ Location of property where plants were collected (name nearby crossroads).

☐ Portion of planted area that the sample represents. Include a map if necessary.

☐ Your observations of previous problems and of when the present problem was first detected.

SHIPPING

☐ Contact the person or laboratory that will receive samples to determine the best method of shipping and to inform them that samples will be arriving.

☐ Pack samples in a sturdy, well-insulated container to prevent crushing or heat damage.

☐ Mark the package clearly and request that the shipper keep it under cool conditions.

☐ Ship packages early in the week so they will arrive before a weekend.

vascular system for damage. Be sure to dig up some infected plants and examine the root system.

Check the distribution of disease symptoms throughout a population of plants. Do many individuals show signs of infection, or only a few? Are the diseased plants distributed evenly, or are they confined to one or more specific locations? Try to associate the distribution of infection with conditions that would favor one disease over another. These factors may provide valuable clues to the identity of the pathogen.

Fungi

Fungi encompass a diverse group of primitive organisms. They absorb nutrients from organic sources around them, such as living or dead plant material. Most fungi live off of dead organic matter and most are considered beneficial because they help break down organic materials and build up soil fertility. Some fungi cause problems, however, invading structural wood and other building materials. Fungi also cause a number of plant diseases as well as animal diseases such as ringworm and athlete's foot. Certain species live off of both dead and living plant material. This last group includes many of the major plant pest species.

Important Characteristics. Most plant fungi have a vegetative body, called the mycelium, which consists of tiny filamentous strands called hyphae. These hyphae often grow through the tissues of an infected host plant and provide the fungus a means for absorbing nutrients. Fungal reproduction is primarily by means of spores. Many species produce more than one type of spore. Some spore types may act as resting structures to carry the fungus through periods of adverse conditions, such as hot, dry weather. Other types of spores are responsible for the ongoing secondary spread of the pest organism.

Disease Symptoms. Plants infected by fungi exhibit any of a variety of symptoms. These symptoms will help you identify the fungal organism. Powdery mildew, downy mildew, root and stem rot, sooty mold, and slime molds are types of fungi.

How Fungi Spread. Wind, rain, irrigation water, propagation (through both grafting and pollenization), seeds, insects, people, equipment, and cultural practices all can spread fungal spores. Any practice that moves infected plant material from one location to another can also spread fungal spores. Roots of noninfected plants may grow into areas where fungal organisms are present and so become infected. Leaves and other aboveground plant parts that touch other plants may aid in spreading infections.

Fungal Infections. Fungi enter plant tissues through wounds and natural openings and by penetrating the outer surface of the plant host. Certain environmental conditions can promote or accelerate fungal growth and infection rates. These often include high humidity or the combined presence of water and warm temperatures.

Bacteria

Pathogenic bacteria are small, usually single-celled organisms, although *Streptomyces* bacteria are composed of many cells attached end-to-end. You cannot see bacteria without a powerful microscope because they are so small. Plant-infecting bacteria are rod-shaped and most have threadlike structures called flagellae that propel them through liquids. Most bacteria require warm, moist conditions if they are to multiply. Because of it's semi-arid climate and lack of significant summer rainfall, only a few bacterial diseases are serious threats to plants in California.

Types of Bacteria. Bacterial species that are capable of infecting plants and causing disease symptoms in

California may belong to any of a number of genera, including *Pseudomonas, Xanthomonas, Agrobacterium, Erwinia, Corynebacterium, Streptomyces,* and *Xylella.*

How Bacteria Spread. Anything that moves and encounters bacteria has the potential to spread them to new areas. This includes farm equipment, rainwater or irrigation water, plant material, seeds, birds, insects, nematodes, and people.

Bacterial Infections. Bacteria invade plant tissues through natural openings and through wounds. They must enter the plant's tissues in order to start an infection. Bacterial diseases show up as wilts, shoot blights, cankers, galls, leaf spots, soft rots, scabs, and systemic disorders. Stone fruits are susceptible to bacterial blight infections. Oleander and olive are susceptible to both bacterial gall and bacterial scorch. Liquidambar and ornamental plum are susceptible to bacterial scorch.

Viruses

Viruses are very small organisms that multiply in living cells to produce disease symptoms. Viruses infect plants and animals. Most viruses can survive for only a short period outside of host plant or animal cells. Viruses alter the chemical activity (metabolism) within host cells, and these changes constitute viral disease.

Important Characteristics. Viruses have a variety of shapes depending on the type, but they are so small that they can only be seen with an electron microscope. Structurally, viruses are much simpler than most other living organisms. A virus invades a cell of a plant or animal and then uses its own genetic information to alter the cell. Infected cells then produce more viruses rather than their usual proteins or nucleic acid.

Symptoms of Viral Diseases. A common viral symptom is a change in coloration, most noticeable in the leaves, and often seen as a mosaic pattern of light and dark blotches. Flowers can also exhibit these color variations. Sometimes the leaf's veins become pale, producing a netlike appearance. *Russeting* is a color change formed by corky, netlike blotches that appear on fruit. Viruses can produce malformations or abnormal growth in many parts of infected plants, including flowers, leaves, fruits, stems, or roots. Some viruses cause parts of infected plants to die, a condition called necrosis (similar symptoms appear when a plant has a fungal infection). Severe necrosis results in the death of the entire plant.

How Viruses Spread. The most common way for viruses to spread on plants is through pruning, grafting, and propagation. Mites and several types of plant-feeding insects such as aphids, thrips, leafhoppers, and whiteflies also carry viruses from plant to plant. The seeds of infected plants transmit some viruses to new areas. Most viruses usually are transmitted in only one of these ways, although some viruses may be transmitted by more than one method.

Abiotic Disorders

Abiotic disorders in plants are problems brought on by external factors such as weather, too little or too much water, nutrient imbalances, chemicals, and other causes.

Important Characteristics. Different plants have different degrees of susceptibility to the effects of abiotic conditions. A plant's susceptibility often depends on a combination of factors including its genetic makeup, current nutritional state, and certain climatic factors. Familiar signs of abiotic disorders in plants include

- poor growth
- yellow, discolored, or partially dead foliage
- deformed or stunted leaves, shoots, and flowers
- deformed or rotten fruit
- dead or dying plants or leaves

Affected plants exhibit symptoms

similar to those caused by diseases, but in this case they are the result of factors such as water stress or excess, nutrient deficiency or toxicity, or salt buildup. In addition, prolonged wind and other weather conditions or exposure to toxic materials such as air pollutants or certain pesticides can produce plant disorders (Table 1-2). Disorders caused by abiotic factors are a result of an organism's response to the conditions in its environment. These cannot be transmitted from one plant to another. Pesticides do not correct or prevent abiotic disorders, but in some cases may cause them. For more information on abiotic disorders, see *Abiotic Disorders of Landscape Plants: A Diagnostic Guide* (UC ANR Publication 3420).

TABLE 1-2.

Soil, Climatic, and Other Factors Contributing to Plant Growth Problems.

ENVIRONMENTAL CONDITION OR OTHER PHYSICAL FACTOR	EFFECT
LIGHT CONDITIONS	
Too little light	Causes yellowing of leaves and elongation of stems. Plants become spindly. May cause twig or shoot dieback.
Too much light	Causes scalding or russeting of leaves; sunburn.
NUTRIENT CONDITIONS	
Nutrient deficiencies	Cause leaf discoloration, leaf spotting, deformities of leaves and fruit; reduce growth and shorten life of plant; plant is unable to tolerate other stresses; causes leaf drop.
Nutrient excess	Causes leaf burn or scald.
TEMPERATURE CONDITIONS	
Cool soil temperature	Slows or shuts down water and nutrient uptake, resulting in wilting or nutrient deficiency symptoms even when soil contains adequate amounts of water and nutrients.
Cool temperature	May reduce pollination or slow fertilization.
Freezing	May cause dieback, leaf scorch, frost nip, bark split. Often causes separation of the epidermis from underlying tissues.
High temperatures	Causes sunburn of foliage. Results in sun scald, heat cankers, sunken lesions, pit burn, blossom end rot, and water core.
Inadequate chilling	Delays or interferes with normal breaking of dormancy; produces irregular bloom.
WATER CONDITIONS	
Too little water	Results in water stress, wilting, poor plant growth, leaf or fruit drop, leaf scorch, and nutrient deficiencies. Leaves may have dull coloration and may roll at edges.
Too much water	Lowers oxygen available to roots; may cause flower, fruit, or leaf drop; lowers soil temperature; may kill roots. Plants may be more susceptible to soil fungi.
OTHER PHYSICAL FACTORS AFFECTING PLANTS	
Air pollution	Causes spotting and damaged foliage, discoloration, reduced growth, and dieback.
Fire damage	Injures bark and nutrient transport system of plant; results in stunting, poor growth.
Hail and rain	Causes physical injury to leaves, resulting in spotting; leaf drop.
Improper cultural practices	Cause physical injury to roots or trunks of plants and reduce water and nutrient uptake; provide an entry point for disease infection and insect infestation.
Pesticide injury	Causes leaf spotting, burning, or scorching; leaf drop; leaf deformity.
Potbound root system	Causes problems similar to those caused by soil compaction; plant exhibits weak growth, yellowed foliage, and may suffer leaf drop and twig dieback.
Soil compaction	Limits root growth and nutrient availability, resulting in problems associated with water stress or excess water and nutrient deficiencies.
Wind	Desiccates leaves, resulting in spotting, burning, or scorching; destruction of blossoms.

CHAPTER 1: REVIEW QUESTIONS

1. Which of the following is not a characteristic of weeds?

☐ a. some weeds are poisonous to people or animals
☐ b. weeds interfere with the safety or use of landscaped areas
☐ c. most weeds produce large quantities of seeds, even under adverse conditions
☐ d. weeds enhance the growth of landscape plants

2. Perennial weeds are generally the most difficult to control because they

☐ a. produce more seeds than other types of plants
☐ b. can reproduce and spread from storage organs such as rhizomes and tubers
☐ c. have seeds that remain viable longer than those of annual and biennial plants
☐ d. have hairy leaves that repel herbicides

3. Which of the following insect orders includes species known to transmit some plant diseases?

☐ a. Orthoptera (crickets and grasshoppers)
☐ b. Thysanura (firebrats and silverfish)
☐ c. Blattodea (cockroaches)
☐ d. Thysanoptera (thrips)

4. Which feature characterizes vertebrates?

☐ a. never lays eggs
☐ b. has fur
☐ c. has backbone
☐ d. all of the above

5. Which of the following orders of insects includes species that are pests of living plants?

☐ a. Isoptera (termites)
☐ b. Thysanura (firebrats and silverfish)
☐ c. Blattodea (cockroaches)
☐ d. Heteroptera (true bugs)

6. The immature stages of an insect are known as

☐ a. instars
☐ b. metamorphoses
☐ c. juveniles
☐ d. pre-adults

7. Which of the following groups of insect orders include both pests and beneficials that attack pest insect species?

☐ a. Hymenoptera (wasps and bees), Thysanoptera (thrips), and Heteroptera (true bugs)
☐ b. Thysanura (firebrats and silverfish), Blattodea (cockroaches), and Dermaptera (earwigs)
☐ c. Isoptera (termites), Mallophaga (chewing lice), and Anoplura (sucking lice)
☐ d. Homoptera (aphids and whiteflies), Lepidoptera (moths and butterflies), and Siphonaptera (fleas)

8. Which of the following insects suck plant juices out of plant cells?

☐ a. Dermaptera (earwigs)
☐ b. Homoptera (aphids and whiteflies)
☐ c. Lepidoptera (moths and butterflies)
☐ d. Orthoptera (crickets and locusts)

9. Which part of a plant do nematodes most commonly attack?

☐ a. fruit
☐ b. roots and other underground plant parts
☐ c. germinating seeds
☐ d. flowers

10. Which of the following types of plant disorders cannot be transmitted from plant to plant?

☐ a. bacterial diseases
☐ b. fungal diseases
☐ c. viruses
☐ d. abiotic problems

11. Ants belong to the insect order

☐ a. Coleoptera
☐ b. Diptera
☐ c. Hymenoptera
☐ d. Lepidoptera

12. Which of the following insects have chewing mouth parts?

☐ a. caterpillars
☐ b. true bugs
☐ c. aphids
☐ d. thrips

13. Vertebrates include which of the following types of animals?

☐ a. spiders and ticks
☐ b. snails and slugs
☐ c. birds and mice
☐ d. butterflies and beetles

14. Which of the following animals are arthropods?

☐ a. fish
☐ b. slugs
☐ c. rats
☐ d. mites

15. Which of the following pests would cause leaf stippling?

☐ a. fungus disease
☐ b. beetle larvae
☐ c. mites
☐ d. caterpillars

16. An insect that passes through complete metamorphosis would have which of the following life stages?

☐ a. egg, nymph, adult
☐ b. egg, larva, adult
☐ c. egg, larva, nymph, adult
☐ d. egg, larva, pupa, adult

17. Powdery mildew on plants is caused by

☐ a. insect feeding damage
☐ b. fungus
☐ c. cultural practices
☐ d. high temperatures

18. An abiotic disorder of a plant can be caused by

☐ a. insects or mites
☐ b. fungal organisms
☐ c. irrigation problems

2 Managing Landscape Pests

A SUCCESSFUL PEST management program in a landscaped, turf, or interiorscape setting involves choosing the right prevention or management methods. It also involves knowledge of how to use these methods to reduce pest populations to acceptable levels. Although this study guide is written to help you understand how to use pesticides properly and safely to manage turf, interiorscape, and landscape pests, it is also intended to make you more aware of a wide variety of pest management methods and to help you understand how these can be incorporated into an integrated pest management (IPM) program.

Besides knowing what the pest is, you must also understand its relative importance. On plants, there may be key pests, occasional pests, and secondary pests.

Key pests are those that will cause major problems unless you successfully manage them. These may be plant mites, insects such as aphids, or plant diseases. Key pests usually require regular management efforts within a pest management program.

Occasional pests are pests that do not cause frequent damage. Their populations may build up to damaging levels from time to time due to life cycle fluctuations, environmental conditions, improper uses of pesticides, or changes in food supply. For instance, ant populations build up when food such as honeydew produced by aphids or other honeydew-producing insects is abundant (Figure 2-1).

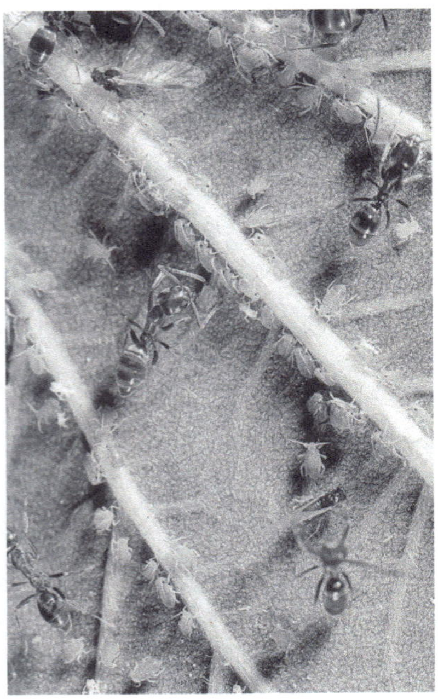

FIGURE 2-1.

Sometimes the presence of one pest will promote other pests. These ants are feeding on the honeydew excreted by aphids infesting this plant.

Secondary pests are those that may cause problems once you eliminate one or more key pests. For example, some weed species appear only when you have successfully managed other weeds. Many insect and mite pests become problems when insecticides applied to manage other pests kill off the natural enemy species that had been keeping the secondary pests below problem levels.

APPROACHES TO PEST MANAGEMENT

Total eradication of a pest from an area is neither practical nor desirable, especially in an outdoor setting, and fortunately it is not often necessary. Eradication may work in buildings or other small, confined spaces, but in turf and landscaped areas eradication efforts will have limited success. This is because most pests can easily migrate back in from untreated surrounding areas.

Most management efforts are intended to lower pest populations rather than eliminate them. Through the use of cultural practices, natural enemies, pesticides, and other pest management practices (see below), you can keep pests in landscape, turf, and interiorscape areas at tolerable levels and minimize plant damage. Sometimes when pest populations have been reduced, the natural enemies that are already present in the area are sufficient to keep them under control.

Preventing Damage

Often there are economical, environmentally sound ways to prevent pest buildup and damage in turf and other plantings, methods that do not rely on the use of pesticides. Such techniques include planting weed- and disease-free turf seed and growing plant varieties that resist diseases or insects under local conditions (Figure 2-2). When planting or replanting an area, carefully inspect all new plant material and reject it if it has any pests. In addition, keep plants adequately watered and fertilized so they will be less susceptible to diseases or pests.

- observe local, state, and federal regulations
- evaluate the success of your pest management efforts

Non-Chemical Pest Management Methods

Once you anticipate or identify pests, begin planning how you are going to manage them and how you can work them into an integrated pest management program. Here are some management methods:

FIGURE 2-2.

Use pest-resistant plants and then water and fertilize them properly to reduce pest problems.

SETTING UP YOUR PEST MANAGEMENT PROGRAM

To begin managing an identified pest, you must

- know what management methods are available
- evaluate the benefits and risks of each method
- select methods that are most effective and least harmful to people and the environment
- integrate several methods whenever possible
- use the methods correctly

Biological Control

Most insect and mite pests have one or more natural enemies. These might be predatory or parasitic insects or mites, spiders, birds, or other animals (Figure 2-3). Biological control programs take advantage of the ability of natural enemies to control pests. In some cases, you may be able to modify the environment in a way that increases the population of natural enemies, for

FIGURE 2-3.

Natural enemies are important components of an integrated pest management program. Here, a convergent lady beetle eats an aphid.

FIGURE 2-4.

Physical barriers are a good way to keep some pests (such as deer and other large animals) out of planted areas.

instance by planting vegetation that attracts them. In addition, you might be able to introduce additional natural enemies (a practice called *augmentation*) that you can purchase from commercial sources.

Naturally Occurring Control. Preservation of existing natural enemy populations can be one of your most economical ways to manage some pests. You can accomplish this by avoiding cultural practices or indiscriminate pesticide and fertilizer use that could destroy the natural enemies. If you need to use pesticides as part of your management program, select types that are less toxic to natural enemies.

Augmentation. Augmentation involves the release of certain natural enemies (usually parasitic or predatory insects or mites) into an area when the target pest crops up. Some of these natural enemies may already occur naturally in the area, but their populations may be too low to control the target pest. Natural enemy populations are not sustainable in an indoor environment, so in these areas augmentation is a necessary and important tool.

Commercial insectaries raise many different types of natural enemies for sale. These include predatory mites to control plant-infesting spider mites as well as parasitic wasps and lacewings to control various insect pests. Lady beetles are useful for controlling certain interiorscape pests.

Mechanical and Cultural Control

Another important way to manage certain pests is to use mechanical and cultural control methods. You can use mechanical devices and cultural techniques either separately or together to

- exclude or trap pests
- destroy pests
- alter pests' life cycles
- encourage or augment natural enemies
- change the environment to make it unsuitable for pest survival
- improve growing conditions as a way to reduce stress on host plants

The following sections describe several mechanical and cultural control methods that you can use to manage pests on plants.

Exclusion. Exclusion is the use of barriers to prevent pests from getting into an area. Fences and ditches make effective barriers against many animals (Figure 2-4). Sticky material painted onto the trunks of trees or shrubs, posts, wires, and other objects can block crawling insects (Figure 2-5). By setting pots and planters in pans filled with water, you can keep crawling pests such as ants away.

Trapping. Traps physically catch pests. Some (most types of gopher trap, for example) kill the animals. Traps can take the form of a mechanical device or just a sticky surface (Figure 2-6).

Cultivation. Cultivation is one of the most important methods for weed control. It also helps control some insects and other soil-inhabiting pests. Devices such as rotary tillers mechanically destroy certain weeds or disrupt their growth (breaking up the rooting structures of certain types of weeds will actually promote the spread of these plants, so caution is advised whenever using rotary tillers). Cultivation also alters and in some cases destroys the conditions suitable for the survival of some soil-inhabiting microorganisms and insects.

Other Cultural Practices. Many other cultural practices influence the survival of pests. Strong, healthy plants that are correctly watered, fertilized, and pruned are most resistant to pest infestation. In turf, proper mowing, irrigation, aeration, and fertilization all are important ways to prevent pest buildup and damage (Figure 2-7).

In some areas, mulching with plastic tarps, straw, or other materials, flooding, solarization, and flaming are effective weed management methods (Figure 2-8). Solarization involves

FIGURE 2-5.

A sticky substance spread onto the trunks of shrubs and trees will keep some pests, such as ants, away.

FIGURE 2-6.

Traps—such as this tunnel trap set to trap tree squirrels—are effective tools for managing certain mammals.

covering moist soil with clear plastic film during warm sunny weather in order to raise the temperature of the soil, thus killing weeds, weed seeds, and some pathogens. Flaming involves exposing weeds to high-temperature flames from a torch-like burner.

Sanitation

Sanitation reduces pest outbreaks in many locations. Sanitation, or *source reduction,* involves eliminating food, water, shelter, or other factors that are important to the pest's survival. This includes removal of dead plants and inoculum-filled plant debris. Sanitation also includes removing weeds that harbor pest insects or rodents and removing or mowing weeds before they produce seed.

FIGURE 2-7.

In turf, proper mowing, irrigation, aeration, and fertilization all are important ways to prevent pest buildup and damage.

FIGURE 2-8.

The use of a mulch, such as these stones applied over plastic, is an effective way to suppress weeds.

Host Resistance

Some problems can be prevented through the use of certain plant varieties that are naturally resistant to insects and plant pathogens. Selecting plants that resist specific pest problems in your area is an effective preventive method of pest management.

Integrated Pest Management

Integrated pest management is an ecosystem-based strategy that focuses on long-term prevention of pests or their damage through a combination of techniques such as biological control, habitat manipulation, modification of cultural practices, and use of resistant varieties. Pesticides are used only after monitoring indicates they are needed according to established guidelines, and treatments are made with the goal of removing only the target organism. Pest management materials are selected and applied in a manner that minimizes risks to human health, beneficial and nontarget organisms, and the environment.

FIGURE 2-9.

Pesticides often have a key role in managing pests and occasionally may be the only management method available to you.

Pesticides

Pesticides often have a key role in managing pests and occasionally may be the only management method available to you (Figure 2-9). Major benefits associated with the use of pesticides include

- effectiveness
- speed and ease of pest control
- reasonable cost (in many instances)

Pest damage usually stops within a few hours (mites and insects) to a few days (weeds) after pesticide application. Most fungicides only provide immediate, short-term protection against infection by disease-causing organisms, although certain materials can control diseases in already-infected plants.

According to California regulations, a pesticide is defined as "... any substance, or mixture of substances which is intended to be used for defoliating plants, regulating plant growth, or for preventing, destroying, repelling, or mitigating any pest ... which may infest or be detrimental to vegetation, man, animals, or households, or be

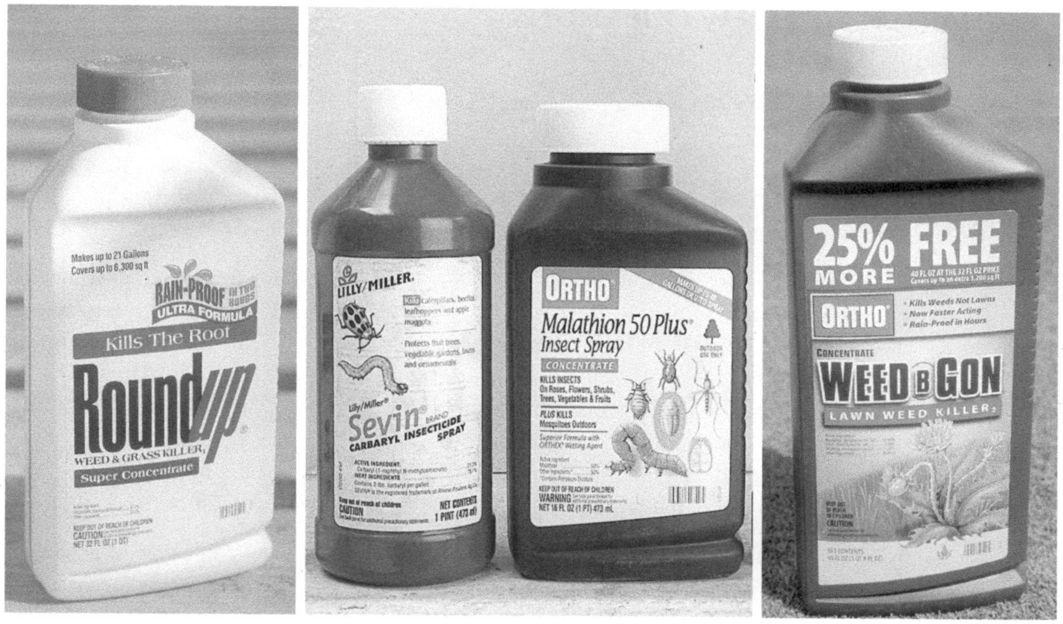

FIGURE 2-10.

Many different types of pesticide are used in landscaped areas, on turf, and in interiorscapes. These may include herbicides, insecticides, fungicides, rodenticides, nematicides, and miticides.

present in any agricultural or nonagricultural environment whatsoever." Herbicides, insecticides, fungicides, rodenticides, nematicides, and miticides are common types of pesticide (Figure 2-10). The next chapter describes pesticide types and ways they are classified.

You must use pesticides *effectively* as well as safely. The results of a pesticide application should be worth any financial or labor investment by reducing health hazards to people and improving health, appearance, and growth of the plants (Figure 2-11).

FIGURE 2-11.

Use pesticides effectively as well as safely. Be sure use is justified by the presence of pests at damaging levels.

CHAPTER 2: REVIEW QUESTIONS

1. Efforts to completely eliminate a pest species are known as

☐ a. eradication
☐ b. suppression
☐ c. sanitation
☐ d. biological control

2. A key pest is one that

☐ a. causes damage only after other pests have been eliminated
☐ b. hardly ever needs to be controlled by pesticides
☐ c. causes damage on a regular basis unless you successfully control it
☐ d. becomes a pest once in a while

3. Applying a preemergence herbicide to an area where weed seeds are present is an example of a ______________ use of a pesticide.

☐ a. preventive
☐ b. nonselective
☐ c. augmentative
☐ d. casual

4. Suppressing a pest population involves

☐ a. completely eliminating the pest
☐ b. treating to control a pest before it appears
☐ c. keeping a pest population below a certain level
☐ d. using only non-chemical control techniques

5. Which of the following is an example of cultural and mechanical control?

☐ a. mowing weeds
☐ b. applying an herbicide
☐ c. releasing imported natural enemies
☐ d. fumigating the area

6. The use of barriers such as screens, fences, and cloth mesh is known as

☐ a. eradication
☐ b. elimination
☐ c. ecology
☐ d. exclusion

7. Sanitation is an important method for controlling

☐ a. rodents
☐ b. abiotic disorders
☐ c. upsurges in natural enemy populations
☐ d. beneficial insects

8. Integrated pest management is an important pest control strategy because it

☐ a. focuses on long-term prevention of pests or their damage
☐ b. completely eliminates pests
☐ c. consistently controls pests without the use of pesticides
☐ d. eliminates the need for constant pest monitoring

9. What type of pest management practice does inspecting plants for insects and diseases before planting them involve?

☐ a. biological control
☐ b. preventative control
☐ c. cultural control
☐ d. mechanical control

10. Removing weeds before they produce seed is what type of pest control?

☐ a. augmentive
☐ b. biological
☐ c. sanitation
☐ d. exclusion

11. Growing plants that resist insects would be what type of pest control practice?

☐ a. sanitation
☐ b. biological
☐ c. exclusion
☐ d. prevention

3 Pesticides

All pesticides are to a certain degree toxic to at least one type—and for many pesticides, several types—of living organism. If you use them improperly for turf, landscape, or interiorscape pest control, you risk harming yourself, others, the plants you are protecting, or the surrounding environment. Therefore the selection, use, and handling of pesticides, including over-the-counter products intended for homeowner use (Sidebar 6), requires that you exercise special skills and receive appropriate training.

When you select a pesticide to manage a landscape, turf, or interiorscape pest, it is useful to learn how the chemical works. Review the following sections for this information. Also, use this chapter to find out how different pesticide formulation types react with target pests, nontarget organisms, the plants, and the environment. You will find more detailed information about pesticides in Chapter 3 of *The Safe and Effective Use of Pesticides, Second Edition.*

PESTICIDE TOXICITY

Toxicity, just like color or odor, is one of the qualities used to describe hazardous chemicals such as pesticides. *Toxicity* is the ability of a pesticide to injure a person (or animal), while *hazard* is the potential a pesticide has to cause the injury. For example, two different pesticides may have similar toxicities or ability to cause injury, but while one may produce vapors that you can easily inhale, the other may not produce vapors. In this case, the first pesticide would be more hazardous. Not all pesticides share the same degree of hazard; some are more toxic than others.

You would not normally use highly toxic pesticides for landscape, turf, or interiorscape pest control because of the serious hazards they pose. Any pesticide will have some level of hazard, however, so you need to know about and use the safe handling procedures for products used on turf and in landscaped areas. Safe handling procedures help to protect you as well as children, adults, and domestic animals who may use the treated area. Labels on pesticide products describe specific hazards and the precautions you must take to avoid them.

Animal Testing

One way scientists determine the toxicity of a pesticide is by testing it on laboratory animals and observing the results. Through this process, researchers find the maximum dose that test organisms can tolerate before being injured. Results from tests of this type help predict hazards to people and nontarget organisms. Animal testing also provides information on a pesticide's modes of action.

During some tests, researchers feed laboratory animals small daily sublethal doses of a pesticide. Such studies help them establish a *no observable effect level* (NOEL). Such studies also give information on the long-term *(chronic)* health effects of pesticide exposure and this information can be used to estimate the effects on people.
For instance, tests on laboratory

SIDEBAR 6

Pesticide Products Labeled for Homeowner Use

Many types of pesticides sold in retail stores are intended primarily for use by homeowners. These can be purchased in small quantities and may be available in ready-to-use (RTU) formulations. This makes them convenient to use for small applications.

Maintenance gardeners and interiorscapers find these over-the-counter RTU products useful for spot treatments because

- ☐ they are convenient for carrying in service vehicles
- ☐ no mixing is necessary
- ☐ no special application equipment is required
- ☐ they eliminate the need to clean application equipment after use
- ☐ empty containers can be disposed of easily

The convenience and availability of these types of pesticide products may make it easy to forget that they are still pesticides and are regulated by state and federal laws. This means that they must always be used according to the label instructions on the container. In general, the regulations that protect employees, including training, use of personal protective equipment (PPE), and record keeping, still apply when employees are using these RTU products, unless employee use of these products is not significantly greater than the typical homeowner use of the products (one or two RTU containers in any given year).

animals help determine a pesticide's potential for causing sterility, birth defects, cancer, or other health problems.

Factors That Influence Pesticide Toxicity

When you apply a pesticide, certain conditions increase or decrease its toxicity and affect its ability to control a pest. These include temperature, humidity, and exposure to sunlight, wind, or rain. Also, some plants or animals respond differently to a given pesticide. In addition, the age, life stage, and general health of a pest may influence how well the pesticide works. For instance, weeds that are water-stressed or are not growing vigorously may be more tolerant to some herbicides.

Once applied, the chemicals in a pesticide usually break down into different compounds. These new compounds may be less or more toxic than the original pesticide. The time it takes for half of the pesticide to break down in the environment, or change from its original form, is its half-life. Soil microbes, ultraviolet light, temperature, quality of the water used in mixing, or impurities combined with the pesticide can lengthen or shorten its half-life. Sometimes impurities that get into pesticide products during manufacture, formulation, storage, or while you are mixing them also affect the half-life. Mixing one pesticide product with another may also increase or decrease toxicity and alter the half-life.

Persistent pesticides are those that remain in their active state in the environment for long periods. These more stable chemicals do not break down quickly.

Pesticide Toxicity Categories

Federal regulations place each pesticide into one of three categories based on its toxicity and potential to injure people or harm the environment. The measure of toxicity for pesticides is the lethal dose that will kill 50 percent of the test animal population—known as the LD_{50}—as shown in Figure 3-1. This dose represents the weight of toxic material in milligrams per kilogram of body weight of the test animal (mg/kg). The lower the value of the LD_{50}, the more toxic a pesticide is. Toxicity categories do not provide information about possible long-term health effects of the pesticides, however. Pesticide labels indicate one of the three toxicity categories by displaying the following signal words:

- *Danger or Danger Poison*—Category I
- *Warning*—Category II
- *Caution*—Category III

Label requirements and legal requirements apply to all three pesticide categories. For example, in California, you must be a certified applicator to get a permit from the county agricultural commissioner's office to buy, possess, and use certain pesticides that are classified as restricted-use pesticides or restricted materials, regardless of their category ranking.

Category I *Danger Poison* and *Danger* Pesticides

Category I *Danger Poison* and *Danger* pesticides have either an oral (through the mouth) toxicity LD_{50} of up to 50 mg/kg or a dermal (through the skin) toxicity LD_{50} of up to 200 mg/kg, or their use may result in some type of serious hazard. The pesticides that fall within this LD_{50} toxicity range listed above also have the word *Poison* and a skull and crossbones symbol on the label (Figure 3-2). Category I *Danger Poison* pesticides with the skull and crossbones symbol on their label are the most hazardous because they are the most toxic—a few drops to a

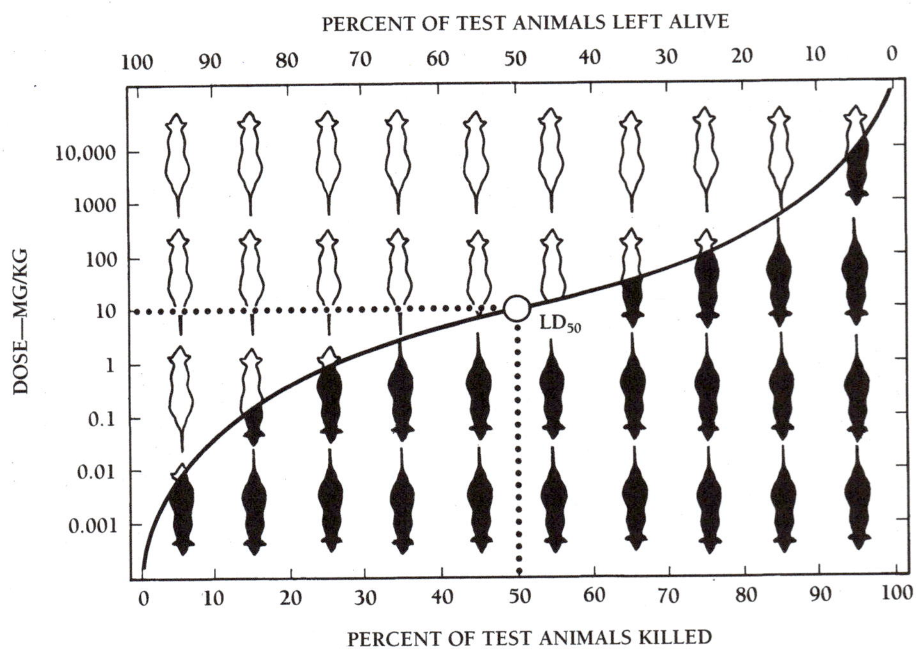

FIGURE 3-1.

Pesticides are often tested on laboratory animals to determine their toxicity. The amount of pesticide that will kill half of a test animal population (illustrated by rats in this drawing) is known as the LD_{50}. "LD" stands for "lethal dose." In this illustration, the LD_{50} is 10 mg of pesticide per kg of body weight of the test rats.

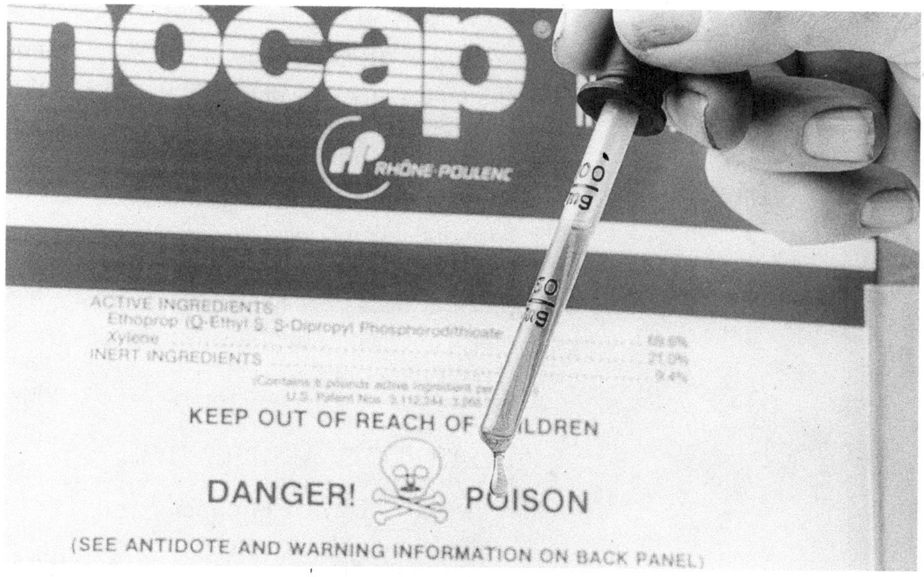

FIGURE 3-2.

Very toxic pesticides will have this skull and crossbones symbol on the label, along with the signal word Poison.

teaspoonful of some of these pesticides could cause death if ingested by a person.

Less-toxic pesticides may still have the signal word *Danger* if they have specific hazards. These hazards include severe skin or eye injury or a particular risk to the environment. For those, the signal word *Danger* appears on the label, but not the word *Poison* or the skull and crossbones symbol.

Category II *Warning* Pesticides

Category II *Warning* pesticides have an oral toxicity LD_{50} of between 50 and 500 mg/kg or a dermal toxicity LD_{50} of between 200 and 2000 mg/kg (Figure 3-3). These pesticides are moderately hazardous. If swallowed, from 1 teaspoon to 1 ounce (6 teaspoons) of many of the pesticides in this group could kill an adult.

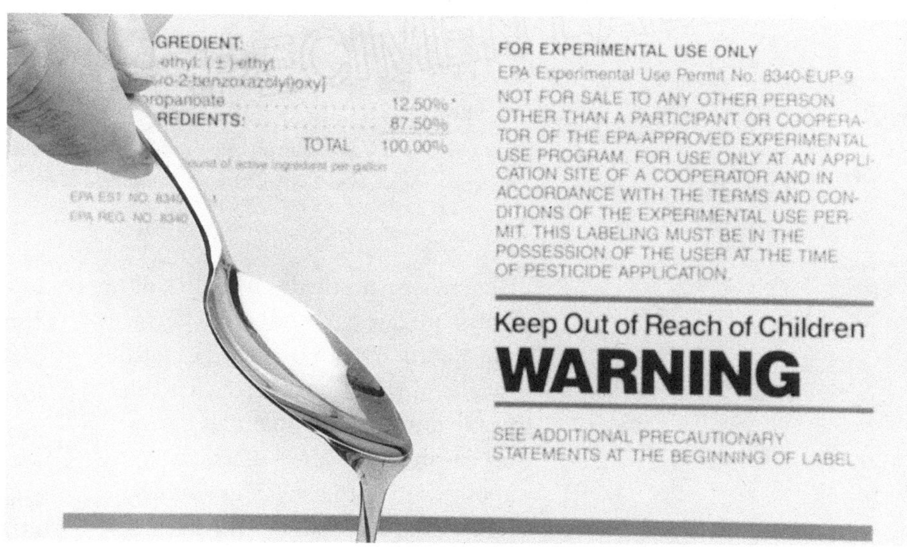

FIGURE 3-3.

The signal word Warning on this label indicates that the pesticide has a moderate toxicity, but still should be handled with care.

Category III *Caution* Pesticides

Category III *Caution* pesticides have an oral toxicity LD_{50} of over 500 mg/kg or a dermal toxicity LD_{50} greater than 2000 mg/kg (Figure 3-4). The signal word *Caution* indicates they are less hazardous than *Danger Poison*, *Danger*, and *Warning* pesticides. An adult would probably need to swallow more than 1 ounce of this type of pesticide to cause serious injury or death.

PESTICIDE CLASSIFICATION

Federal and State registration processes include classification of pesticides as *general-use* or *restricted-use*.

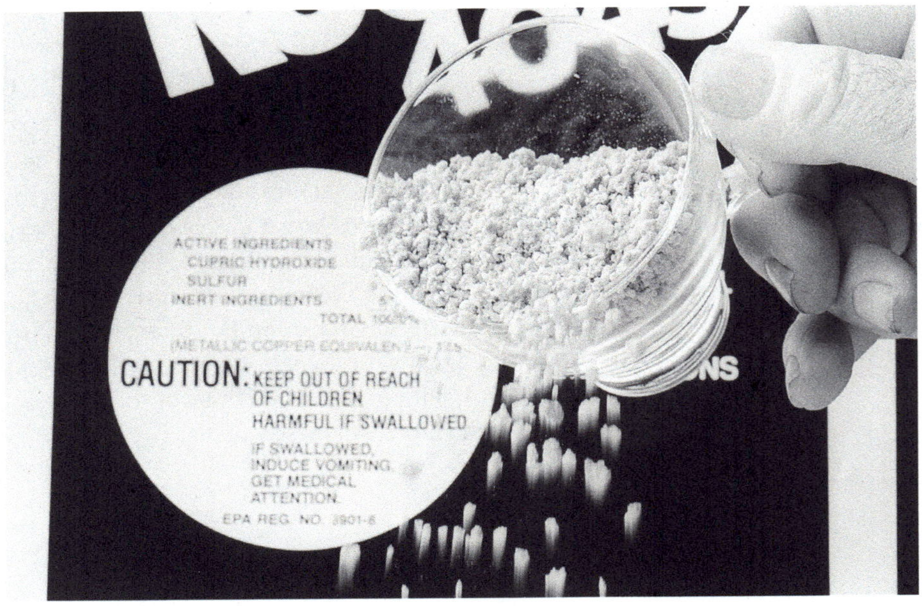

FIGURE 3-4.

The signal word Caution on this label indicates that the pesticide belongs to the least toxic group of materials. All pesticides are hazardous, however, and must be handled with care.

General-use pesticides usually will not cause unreasonable adverse effects on the environment when used according to label instructions. Restricted-use pesticides are those that may cause unreasonable adverse effects to applicators or the environment even when used according to the label. Restricted-use pesticides usually have more regulatory requirements for their use and handling, and users must possess applicator certification and/or a use permit to use them.

PESTICIDE DESIGNATION

Experts designate pesticides according to their function or the type of pests controlled (Table 3-1). For example, insecticides control insects and herbicides control plants. Many plant growth regulators improve growth or fruiting of cultivated plants. Attractants and repellents are pesticides because of their use in pest control.

Some pesticides control more than one group of pests and therefore belong to more than one pesticide class. The compound 2,4-D, for instance, will regulate plant growth when used at low rates, and is therefore classified as a plant growth regulator. It is also effective in controlling broadleaf weeds when applied at higher concentrations, so for these uses it is classified as an herbicide. Fumigants control broad ranges of pest types such as weeds and

TABLE 3-1.

Pesticide Classification Based on Target Pest and Pesticide Function.

PESTICIDE TYPE	FUNCTION
acaricide or miticide	mite control
algaecide	algae control
attractant	attract pests or other animals and invertebrates
avicide	bird control
bactericide	bacteria control
defoliant	remove plant foliage
desiccant	remove water from arthropod pests
fungicide	fungal control
growth regulator	regulate plant or animal growth
herbicide	weed control
insecticide	insect control
molluscicide	snail or slug control
nematicide	nematode control
piscicide	fish control
predacide	predator mammal control
repellent	repel pest animals or invertebrates
rodenticide	rodent control
silvicide	tree and woody shrub control

weed seeds, insects and mites, vertebrates, and soil pathogens.

Pesticide Chemical Groups

Pesticides are also grouped according to their chemical origin. This type of grouping often indicates common characteristics such as mode of action, chemical structure, and types of formulations possible. There also may be similarities between environmental persistence and the way related pesticides break down through biological processes.

Special Types of Pesticide

A few pesticides do not belong to a specific chemical group. Some occur naturally, while living organisms such as plants produce others. Pesticide manufacturers may also synthesize these compounds in their laboratories. These special types of pesticides include antibiotics, anticoagulants, botanicals, insect growth regulators, inert dusts, microbials, petroleum oils, pheromones, plant hormones, and soaps. Refer to *The Safe and Effective Use of Pesticides, Second Edition* if you need information about these special types of pesticide.

MODE OF ACTION

The way a pesticide controls a target organism is its *mode of action.* An understanding of modes of action makes it easier for you to choose the right pesticide and helps you predict which pesticide will work best in a particular situation. With this knowledge, if you notice that pests are showing resistance to one pesticide, you can select another with a different mode of action.

Usually, pesticides within a chemical group have the same mode of action on similar pests. They may also have other characteristics in common, such as chemical structure, persistence in the environment, and formulation.

The effect of a pesticide on a pest may be local, systemic, or both. When you use an herbicide, for example, local injury usually involves plant tissues that the herbicide contacts first, such as leaves. Pesticides with systemic action move within the plant after application, traveling to other plant tissues where the intended damage occurs. For instance, some leaf-applied herbicides move through the plant to its growing points (roots and shoots).

The mode of action of some herbicides causes them to destroy weeds by damaging leaf cells. As a result, the plants dry out. Other herbicides interfere with a plant's ability to grow normally or to convert light into food. The mode of action often determines when and how an herbicide is used. For instance, if the herbicide prevents the germinating weed seed from developing (preemergent), you must apply it before seedlings germinate through the soil surface.

Application of systemic insecticides to a host plant sometimes protects it from certain pests by killing these pests when they feed on the plant. For example, plant roots pick up soil-applied systemic insecticides and the plant carries these insecticides in plant fluids to the leaves. This makes those leaves toxic to foliage-feeding insects.

Insecticides can be nerve poisons, muscle poisons, desiccants (causing the insect to lose water), growth regulators, or sterilants. Others have purely physical effects that clog the insect's air passages. Often an insecticide will have more than one mode of action.

Some fungicides are eradicants because they destroy fungi that have already invaded and begun to damage plant tissues. Their mode of action is to slow or stop the growth of the fungal organisms. Other fungicides are protectants and prevent future fungal infections. Their mode of action blocks the fungal organisms from entering the cells of the treated plants.

Factors Influencing Pesticide Effectiveness

Several factors influence how a pest reacts to a pesticide. Two of these are very important:

- The target pest must absorb the pesticide (pesticide uptake).
- The target pest must be in a susceptible life stage.

Pesticide Uptake

Most pesticides act on certain sites inside the tissues of pests, and the pesticide may have to move through other tissues in order to reach these sites. Structural differences, protective coatings, and habits of the pest influence the way a pesticide moves through tissues. The type of pesticide formulation as well as environmental conditions also influence this uptake.

Pesticides with contact activity, such as contact herbicides, pass through the target organism's outer covering (for example, a weed plant's cuticle). Some insecticides and rodenticides are stomach poisons: the target insect or rodent must eat the toxin, which then absorbs through the lining of the pest's mouth or intestines. Many insecticides are both contact and stomach poisons. Other pesticides have fumigant activity and pass as vapors into the tissues of the target plant or animal. The fumigant enters through breathing channels or by passing through skin or cuticle.

Life Stage

The life stage of the pest influences its response to a pesticide. For instance, young weeds generally are more susceptible to certain herbicides than are older weeds. However, some herbicides are most effective on perennial weeds just beginning to flower and less effective on weeds that have not yet begun to flower or have completed their flowering stage.

Insects pass through several life stages including eggs, nymphs or larvae, pupae (in some orders), and adults. Each insect life stage has different susceptibilities to insecticides due to biological and physical characteristics, feeding habits, and the proximity of the pest to the insecticide.

PESTICIDE FORMULATIONS

Pesticide chemicals in their "pure" active ingredient state are rarely suitable for pest control. These concentrated chemicals may not mix well with water or may be chemically unstable, so manufacturers add substances to improve mixing, application, safety, handling, and storage. The final product is a *pesticide formulation.* This formulation may consist of

- the pesticide chemical (active ingredient)
- the carrier, such as an organic solvent or mineral clay (inert ingredient)
- surface-active ingredients, such as stickers and spreaders (inert ingredients)
- other ingredients such as stabilizers, dyes, and chemicals that improve or enhance pesticidal activity (inert ingredients)

Inert ingredients include all of the additives that manufacturers combine with the active ingredient. Inert ingredients do not have pesticidal action, but they still may be hazardous to people or the environment.

To apply a pesticide, you usually need to mix a formulation with water or some other diluent as directed on the pesticide label. Some formulations, however, such as baits, granules, and dusts, are ready for use without additional dilution. Manufacturers also package specialized pesticides, such as some insect sprays and herbicides, in ready-to-use (RTU) formulations.

The pesticide label lists the amount of actual pesticide in a dry formulation as percentage of active ingredient (a.i.). For instance, a 50-W wettable powder contains 50 percent by weight of actual pesticide. With liquid formulations, the label lists the pounds of active ingredient in one gallon of formulated pesticide. For example, a 4-E liquid contains 4 pounds per gallon of the active ingredient.

Some labels indicate a formulation type by letters that follow or are a part of the brand name of the pesticide (Table 3-2). In the examples in the previous paragraph, the "W" represents a wettable powder. The "E" indicates that the pesticide is an emulsifiable concentrate.

Selecting a Formulation

Sometimes you need to choose which of two or more formulations of the same pesticide to control a pest. When possible, make your selection based on the degree of control desired, safety, cost, and other factors such as those listed in Table 3-3. For example, emulsifiable formulations of insecticides usually provide control that is quick but less lasting than that provided by wettable powders. Whenever you have a choice of materials, consider the safety of pesticide applicators, helpers, and the public. Also consider the potential impact on children, domestic animals, and the environment. If natural enemies of the pest are present, select products that will not harm them. Also, take into account the method of application and ease of measurement when selecting a formulation.

Evaluate the habits and growth patterns of the pest and make sure the formulation is suitable for the pest's life stage. Pick a formulation that causes the least impact on the environment. Consider drift, runoff, wind, and rainfall along with soil type and the character of the surrounding area. Cost may also influence your selection. Finally, choose a formulation that is compatible with available application equipment.

TABLE 3-2.

Formulation Types and Their Uses.

TYPE*	USE	COMMENTS
Bait (RB)	Insecticides, rodenticides, avicides	May be an attractive food substance coated on or impregnated with toxic material. Used in bait stations or bait blocks or scattered in safe locations.
Bait block (BB)	Rodenticides	Bait compressed into blocks of different sizes, impregnated with paraffin; especially useful under moist conditions that would normally render the bait useless.
Bait concentrate (CB)	Rodenticides	A liquid concentrate that is diluted with water and applied as drinking water for rodents.
Capsule suspension (CS)	Insecticides	Used for slow, sustained release of active ingredient over time.
Contact liquid or gel (CL)	Insecticides, rodenticides	Liquid or gel, usually applied without dilution.
Contact powder (CP)	Insecticides, rodenticides	Highly visible residues. Must be kept dry. Some types may be used as a tracking powder.
Dry flowable (F or DF)	Insecticides, fungicides, herbicides	Requires agitation during application. May leave visible residues on treated surfaces after drying. Does not penetrate surfaces well.
Dusting powder (DP)	Insecticides	A free-flowing powder suitable for dusting.
Emulsifiable concentrate (EC or EO)	Insecticides, fungicides, herbicides	May damage plants. May cause spotting or staining of treated surfaces.
Flowable (F or SC)	Insecticides, fungicides, herbicides	Requires agitation during application. May leave visible residues on treated surfaces after drying. Does not penetrate surfaces well.
Fumigant (GA or GE)	Insecticides, sometimes used for control of rodents	Must be used in tightly closed area. Requires special application techniques and equipment.
Granular bait (GB)	Insecticides, rodenticides, molluscicides	Special form of bait.
Granule/pellet (GR)	Insecticides, rodenticides	Formulated as bait to be used in bait stations. May have special forms such as encapsulated granule (CG), fine granule (FG), macrogranule (GG), or microgranule (MG).
Impregnate	Insecticides, fungicides	Includes items such as flea collars, pest strips, and special paints. May also include factory-treated fabrics or carpets.
Low-concentrate solution (AE, AL, or AP)	Ready-to-use insecticides	High cost per unit of active ingredient. Very useful in control of insects in interiorscapes. Usually does not cause stains.
Soluble powder (SP)	Insecticides, fungicides, herbicides	Penetrates better than wettable powders or dry flowable formulations.
Suspended concentrate (SC)	Insecticides, fungicides, herbicides	A solid or liquid intended for dilution before use as bait.
Water-soluble concentrate (SL)	Insecticides, fungicides, herbicides, and rodenticides (as drinking solutions)	Penetrates better than wettable powder or dry flowable formulations.
Wettable powder (WP)	Insecticides, fungicides, herbicides	Requires agitation during application. May leave visible residues on treated surfaces after drying. Does not penetrate surfaces well.

**The letters in parentheses that follow the formulation type are the approved abbreviations for each formulation.*

Common Pesticide Formulations

You cannot use most pesticide formulations in their undiluted form. Except for granules, dusts, and ready-to-use (RTUs) formulations, you must dilute and mix pesticide materials with water or some other diluent as directed on the pesticide label before you apply them as liquid sprays.

Manufacturers package some pesticide products in ways that increase their safety or make them easier to use. The water-soluble bag is an example of packaging that increases the handler's safety. The following section describes dry and liquid formulations that you might use in turf, landscape, or interiorscape situations. For more detailed information about these and other pesticide formulation types, consult *The Safe and Effective Use of Pesticides, Second Edition.*

Wettable Powders (W or WP)

Wettable powders do not dissolve in water. Instead, they form a milky suspension. Wettable powders consist of the pesticide and a finely ground dry carrier, usually mineral clay, and should not be confused with dust formulations. Manufacturers combine wettable powder products with other ingredients that improve the powder's ability to remain suspended in water. Most wettable powder formulations are between 15 and 75 percent active ingredient. Wettable powders

- are less damaging to plants than emulsifiable concentrates
- may leave visible residues on sprayed surfaces
- can be inhaled accidentally during handling before they are mixed with water
- are abrasive to application equipment nozzles and pumps

TABLE 3-3.

Comparison of Pesticide Formulations.

FORMULATION	MIXING/ LOADING HAZARDS	PHYTOTOXICITY	EFFECT ON APPLICATION EQUIPMENT	AGITATION REQUIRED	VISIBLE RESIDUES	COMPATIBLE WITH OTHER FORMULATIONS
dry flowables/ water-dispersible granules	safer	safer	abrasive	yes	yes	good
dusts	severe inhalation hazards	safer	—	yes	yes	—
emulsifiable concentrates	spills and splashes	maybe	may affect rubber pump parts	yes	no	fair
flowables	spills and splashes	maybe	may affect rubber pump parts; also abrasive	yes	yes	fair
granules and pellets	safer	safer	—	no	no	—
micro-encapsulated formulations	spills and splashes	safer	—	yes	—	fair
soluble powders	dust inhalation	usually safer	non-abrasive	no	some	fair
solutions	spills and splashes	safer	non-abrasive	no	no	fair
wettable powders	dust inhalation	safer	abrasive	yes	yes	highly

- require constant agitation to prevent the formulation from settling in the spray tank

Dry Flowables or Water-Dispersible Granules (DF or WDG)

Dry flowables (also called *water-dispersible granules*) are similar to wettable powders, but rather than a powder, the formulated pesticide consists of small granules. These must be mixed with water before use. The granules have a higher percentage of active ingredients per unit of weight because the dry flowable formulation allows manufacturers to use less carrier. Dry flowables

- do not have the dust problems associated with wettable powders
- are easier to measure and mix because manufacturers package them in easy-to-pour plastic containers
- require constant agitation during mixing and application
- are abrasive to application equipment

Soluble Powders (S or SP)

A soluble powder is similar to a wettable powder, but rather than being suspended in liquid, the pesticide active ingredient, its carrier, and all other formulation ingredients completely dissolve in water to form a true solution. Once mixed in the spray tank, soluble powders

- require no additional mixing or agitation
- are not abrasive to spray nozzles or pumps

A hazard associated with soluble powders is that they are made up of fine particles that can be inhaled during handling and mixing.

Emulsifiable Concentrates (E or EC)

Many pesticides are not soluble in water but dissolve instead in petroleum solvents. Emulsifiable concentrates are petroleum-soluble pesticides combined with soap-like materials. When you add emulsifiable concentrates to water, they form an emulsion (a milky liquid). After mixing, the emulsified pesticide disperses evenly in the water. The material must be agitated during application to keep the emulsion uniform.

Emulsifiable concentrates are among the most versatile of all formulations and have many applications. They penetrate porous materials such as soil and wood better than wettable powders. Since they are liquid, they pour easily for mixing. On the other hand, emulsifiable concentrates

- spread easily and are difficult to clean up if spilled
- absorb readily into porous protective clothing and leather boots
- pass through the skin more readily than powders do
- may cause serious injury if splashed into the eyes
- are more risky than wettable powders if used on sensitive plants
- contain solvents that may contribute to the deterioration of rubber and plastic hoses and gaskets on the application equipment

Flowables (F)

A flowable formulation combines many of the characteristics of emulsifiable concentrates and wettable powders. Manufacturers use this formulation when the active ingredient is an insoluble solid and will not dissolve in either water or oil. They combine fine-ground pesticide particles with a liquid carrier and emulsifiers to form a concentrated emulsion. Flowables share the qualities of liquid emulsifiable concentrates, even to the point of having similar disadvantages. Flowables

- require agitation to keep them in suspension
- leave visible residues similar to those left by wettable powders
- are subject to spilling and splashing as are emulsifiable concentrates

- contribute to abrasive wear of application equipment nozzles and pumps

Water-Soluble Concentrates or Solutions (L)

Water-soluble concentrates or solutions are formulations that dissolve in water. Water-soluble concentrates

- require no agitation
- are not abrasive to application equipment

Microencapsulated Materials

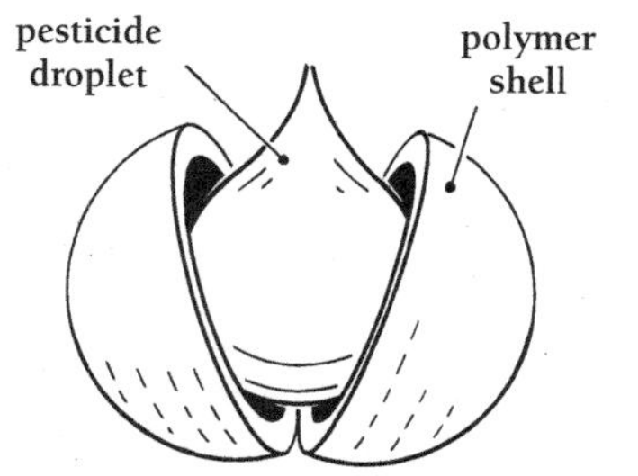

FIGURE 3-5.

Microencapsulated formulations consist of individual droplets of the pesticide active ingredient, each enclosed in a tiny plastic capsule. This often makes the pesticides safer to handle than an equivalent amount of active ingredient in a different formulation and extends the length of their effectiveness.

Manufacturers may coat certain liquid or dry pesticide particles with a plastic material, producing a microencapsulated formulation (Figure 3-5). Microencapsulated pesticides are mixed with water and then sprayed in the same manner as other sprayable formulations. After spraying, the plastic coating breaks down slowly, releasing the active ingredient. There are several advantages to microencapsulated formulations:

- They make highly toxic materials safer for applicators to mix and apply.
- The delayed or slow release of the active ingredient prolongs its effectiveness, allowing for fewer and less-precisely timed applications.
- The pesticide volatilizes more slowly, so less drifts away from the application site.
- They reduce phytotoxicity problems.

Granules (G)

Granules consist of a pesticide and a carrier (such as clay) combined with a binding agent. They range in size between 4 and 80 mesh. The most common formulations are in the range of 15 to 30 mesh. *Mesh* describes the number of wires running in one direction in an inch of screen (Figure 3-6). A larger number indicates more wires and so a finer screen (dusts pass through 80 mesh and coarser screens). The larger size and weight of granules eliminate drift. You can minimize dust and spray mist hazards to the applicator and the environment by using granular formulations. Granular formulations are more persistent in the environment than other formulations because the active ingredient releases more slowly.

You do not mix granules with water or other diluents. Granules can be applied directly to soil to control weeds, nematodes, snails and slugs, and soil insects. In addition, manufacturers formulate some systemic insecticides as granules. Once applied,

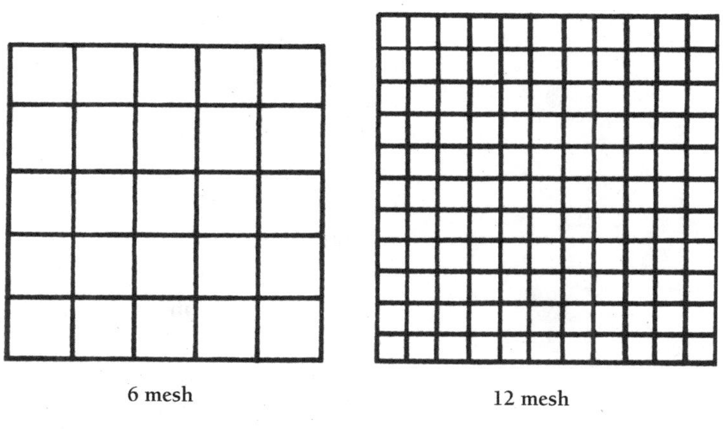

FIGURE 3-6.

The size of dust and granule particles is measured by passing the material through wire screens with different sizes of mesh. The numeric value of a mesh indicates the number of wires running in one direction per inch of screen. The larger the mesh number, the finer the screen. Granules usually range in size between 4 and 80 mesh, while dusts are 80 mesh and finer.

these granules dissolve in the soil so plants can absorb the active ingredients through their roots. These materials control leaf- and stem-feeding insects.

Baits

Baits are pesticides that are combined with food, attractants, or feeding stimulants. Because baits attract target pests to a pesticide, they eliminate the need for widespread pesticide application. Sometimes target pests such as ants carry baits back to their nest-bound young. Baits are used to control slugs, snails, certain insects, and rodents. Dangers associated with baits include their attractiveness to nontarget animals and to children, so tamper-resistant bait stations should be used whenever possible. A bait colored with a dye is less attractive to nontarget organisms.

Fertilizer Combinations

Manufacturers combine some lawn and turf fertilizers with herbicides or insecticides. These combinations are used to control pests while fertilizing the grass, although the best times for fertilizing and pest control application rarely coincide.

PESTICIDE TANK MIXES

Combining two or more pesticides in the same tank and applying them at the same time is convenient and often cost-effective. Only a few pesticide manufacturers sell products as premixed combinations, so usually you must combine the pesticides at mixing time prior to application. When you combine two or more pesticides or a pesticide and a fertilizer, you create a *tank mix*. A common tank mix combines fungicides with insecticides for a dormant spray used on deciduous trees. Another common mix combines two or more herbicides to increase the number of weed species that will be controlled. Some people mix pesticides with micronutrients or fertilizers.

Tank mixes of two or more pesticides save money by reducing the time, labor, and fuel required for multiple applications in certain situations. In these cases, a tank mix reduces equipment wear and decreases labor costs. It may also lessen the degree of turfgrass soil compaction caused by application equipment.

Incompatibility

Incompatibility is a physical condition that prevents pesticides from mixing properly to form a uniform solution or suspension in the spray tank. Precipitation of flakes, crystals, or oily clumps—or severe separation—is unacceptable (Figure 3-7). Such incompatible mixtures may make the pesticide solution unusable, creating a hazardous waste that must be disposed of legally and properly. At the very least, incompatible mixtures clog application equipment and prevent good pesticide coverage.

For more information on tank mixes, incompatibility, and ways to test for and resolve incompatibility problems, see *The Safe and Effective Use of Pesticides, Second Edition.*

ADJUVANTS

Adjuvants are materials that you can add to the spray tank to improve mixing and application qualities or to enhance pesticide performance. Manufacturers formulate pesticides to be suitable to many types of application conditions, but they cannot be formulated for all possible situations. Use adjuvants to customize the formulation to meet specific needs and compensate for local conditions. Adjuvants are used to

- improve the wetting ability of a spray solution (see Figure 3-8)
- control evaporation of spray droplets
- improve the pesticide's ability to stick to plant surfaces so it will not wash off easily

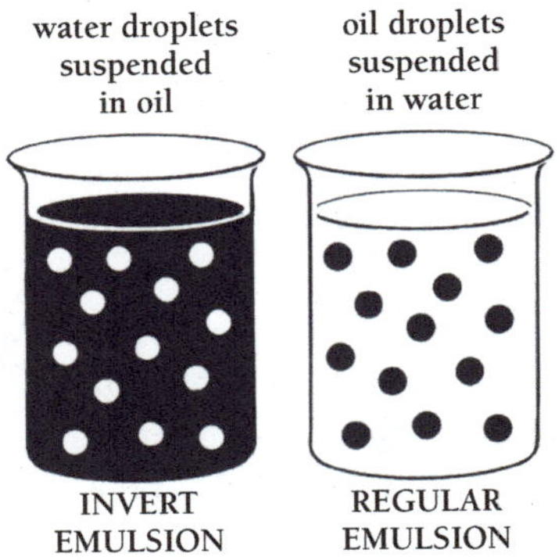

FIGURE 3-7.

Sometimes mixtures of two or more pesticides are incompatible and may separate or curdle. At the very least, this clogs spray equipment, wastes the pesticide material, and may cause an uneven application. It could make the pesticide solution unusable, which would then make it a hazardous waste that would have to be disposed of legally and properly.

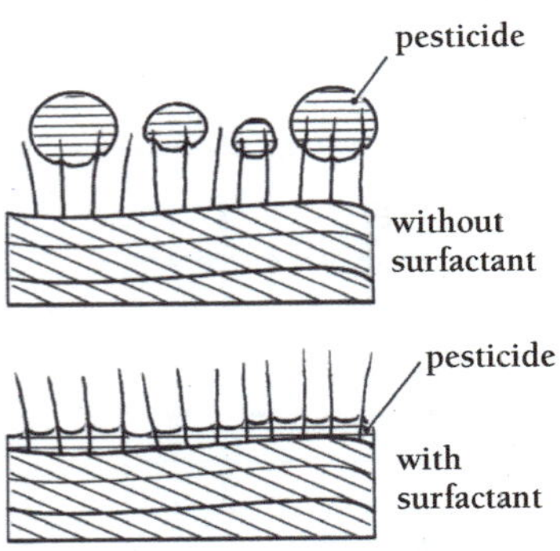

FIGURE 3-8.

Spray droplets will not spread out over waxy or hairy leaf or insect surfaces because of surface tension. Adding a surfactant, however, lowers surface tension and causes the droplets to spread out and contact the cuticle of the leaf or insect.

- improve the penetration of pesticides through plant or insect cuticles
- adjust the pH of spray solutions *(see below)*
- improve spray droplet deposition patterns
- reduce damage to target plants
- correct incompatibility problems
- reduce pesticide drift

Familiarize yourself with adjuvant types so you will understand where and how to use them (Table 3-4). When selecting an adjuvant, know the effect you wish the adjuvant to have. Next, check pesticide and adjuvant labels to make sure you can mix these materials together safely and that they are suitable to the application site, target pest, and application equipment. For a complete discussion of adjuvant types and uses, see *The Safe and Effective Use of Pesticides, Second Edition.*

Adjusting the pH of the Spray Solution

Sometimes water may be too acidic or alkaline to be properly mixed with pesticides. The measure of a substance's acidity or alkalinity is expressed as its pH. A pH level below 7.0 indicates that the substance (in this case, water) is acidic. If the pH is above 7.0, it is alkaline. Many pesticides are unstable in alkaline solutions, but quite stable if the solution is slightly acid. The optimum pH for most pesticides is about 6, although solutions in the range of pH 6 to 7 are usually satisfactory. Some pesticides are most effective when the solution is acidified to a pH of 3 to 3.5.

High pH (alkalinity) often speeds up pesticide breakdown. Sidebar 7 tells how to measure and alter the pH of a spray solution, should this be necessary.

Buffers. A buffer is a substance that can change the pH of a water solution

TABLE 3-4.

Comparison of Adjuvants.

FUNCTION	TYPE OF ADJUVANT											
	Surfactant	Sticker	Spreader-Sticker	Extender	Activator	Compatibility Agent	Buffer	Acidifier	Deposition Aid	Defoamer	Thickener	Attractant
Reduce surface tension	■		■		■							
Improve ability to get into small cracks	■		■									
Increase uptake by target	■		■		■			■				
Improve sticking	■	■	■									
Protect against wash-off/abrasion	■	■	■	■								
Reduce sunlight degradation		■	■	■								
Reduce volatilization	■	■	■									
Increase persistence			■	■					■		■	
Improve mixing						■	■	■				
Lower pH							■	■				
Slow breakdown							■	■				
Reduce drift									■	■	■	
Eliminate foam										■		
Increase viscosity									■		■	
Increase droplet size									■		■	
Attract pests to pesticide												■

to a prescribed level. The buffer keeps this level relatively constant in the solution, even if conditions such as water alkalinity may change.

Acidifiers. An acidifier (also called an *acidulator*) is an acid that neutralizes alkaline solutions and lowers their pH. An acidifier does not have a buffering action like a buffer; that is, it will *not* maintain a steady pH level in the solution if other pH-related conditions change. This being the case, alkaline or acid compounds added to the spray solution *after* the acidifier may change the pH of the solution.

SIDEBAR 7

Testing and Adjusting the pH of Water Used for Mixing Pesticides

The acidity or alkalinity of water (its pH) can be measured with an electronic pH meter, a pH test kit such as those used for testing swimming pool water, or pH test paper available from a chemical supply dealer.

TEST WATER

1. Using a clean container, obtain a sample of water from the same source that will be used to fill the spray tank.

2. Measure exactly 1 pint of this water into a clean quart jar.

3. Check the pH of the water using a pH meter, test kit, or test paper.

    pH 3.5 to 6.0: Water is satisfactory for spraying and short-term (12 to 24 hours') storage of most spray mixtures in the spray tank.

    pH 6.1 to 7.0: Water is adequate for immediate spraying of most pesticides. To prevent loss of effectiveness, do not leave the spray mixture in the tank for more than 1 to 2 hours.

    above pH 7.0: Add buffer or acidifier.

ADJUST PH (JAR TEST)

1. Using a standard eyedropper, add 3 drops of buffer or acidifier to the measured pint of water.

2. Stir well with a clean glass rod or other clean, nonporous utensil.

3. Check pH as above.

4. If further adjustment is needed, add 3 more drops of buffer or acidifier, stir well, and then recheck pH. Repeat until pH is satisfactory. Remember how many times you had to add 3 drops to bring the solution to the proper pH.

ADJUST PH IN SPRAY TANK

1. Before adding pesticides to the sprayer, fill the tank with water.

2. For every 100 gallons of water in the spray tank, add 2 ounces of buffer or acidifier for each time you added 3 drops in the jar test above. Add the buffer or acidifier to the water while agitators are running. If the tank is not equipped with an agitator, stir or mix well.

3. Check the pH of the water in the spray tank to make sure it is correct. Adjust if necessary.

4. Add pesticides to the spray tank.

CHAPTER 3: REVIEW QUESTIONS

1. Which part of the label gives you an indication of the toxicity of the pesticide?
- ☐ a. Statement of Use Classification
- ☐ b. Signal Word
- ☐ c. Directions for Use
- ☐ d. EPA Registration Number

2. The toxicity of a pesticide is usually measured by its
- ☐ a. no observable effect level (NOEL)
- ☐ b. long-term health effects
- ☐ c. LD50
- ☐ d. half-life

3. Which signal word indicates the least hazardous pesticide?
- ☐ a. Poison
- ☐ b. Danger
- ☐ c. Warning
- ☐ d. Caution

4. The toxicity of a pesticide is the
- ☐ a. length of time it remains active in the environment
- ☐ b. potential for reaching groundwater
- ☐ c. boiling point
- ☐ d. capacity to cause injury

5. A post-emergent contact herbicide
- ☐ a. is applied before weeds germinate through the soil surface
- ☐ b. must be translocated in the plant to be effective
- ☐ c. causes injury to any part of the plant it touches
- ☐ d. provides systemic weed protection

6. How much active ingredient would be found in a 20 pound bag of a 25W formulation?
- ☐ a. 4 pounds
- ☐ b. 5 pounds
- ☐ c. 10 pounds
- ☐ d. 20 pounds

7. How much active ingredient would there be in a 5 gallon bottle of a 4EC formulation?
- ☐ a. 4 pounds
- ☐ b. 20 pounds
- ☐ c. 4 quarts
- ☐ d. 20 gallons

8. Which of the following formulations requires no further agitation once it is mixed fully with water?
- ☐ a. emulsifiable concentrate (EC)
- ☐ b. soluble powder (SP)
- ☐ c. wettable powder (W or WP)
- ☐ d. flowable (F)

9. The tendency of ants to take poisoned bait back to the nest
- ☐ a. makes ant bait too dangerous for use around people
- ☐ b. contributes to the effectiveness of ant baits
- ☐ c. can cause ant populations to grow rapidly
- ☐ d. makes ant baits illegal for landscape use

10. The way a pesticide destroys or controls a target organism is its
- ☐ a. half-life
- ☐ b. toxicity
- ☐ c. mode of action
- ☐ d. phytotoxicity

11. If a mixture of two pesticides in the spray tank results in clumping and the spray nozzles clog up, the mixture is
- ☐ a. synergistic
- ☐ b. antagonistic
- ☐ c. surface active
- ☐ d. incompatible

12. The hazard of a pesticide includes its potential to
- ☐ a. cause injury
- ☐ b. control the pest
- ☐ c. damage application equipment
- ☐ d. remain tightly bound to soil particles

13. A pesticide with the signal word "Warning" will generally be
- ☐ a. more hazardous than a pesticide with the signal word "Danger"
- ☐ b. less hazardous than a pesticide with the signal word "Caution"
- ☐ c. about as hazardous as a pesticide with the signal word "Caution"
- ☐ d. less hazardous than a pesticide with the signal word "Danger"

14. A persistent pesticide is one that
- ☐ a. can be stored for up to 5 years
- ☐ b. changes rapidly into different compounds
- ☐ c. is active in the environment for long periods
- ☐ d. easily contaminates application equipment

4 Pesticide and Pest Control Laws and Regulations

STATE AND FEDERAL laws regulate the manufacture, sale, transportation, and use of pesticides. At the national level, the U.S. Environmental Protection Agency (U.S. EPA) is the pesticide regulatory agency. In California, the Department of Pesticide Regulation (DPR) assumes this role and also regulates other types of pest control methods and devices. U.S. EPA's authority is a mandate from the Federal Insecticide, Fungicide, and Rodenticide Act (FIFRA). Originally passed in 1947, this law has undergone several amendments and updates, including the 1992 *Worker Protection Standard*. An important function of FIFRA and the U.S. EPA and DPR regulations is to protect pesticide handlers, workers, and the public from pesticide exposure. See Sidebar 8 for a list of reasons for having pesticide laws and regulations.

Based on FIFRA, U.S. EPA establishes regulations for pesticide registration and labeling and for setting pesticide residue tolerance levels in foods. The regulations also set standards for using restricted-use pesticides: an individual

SIDEBAR 8

Reasons for Pesticide Laws and Regulations

1. To provide for the proper, safe, and efficient use of pesticides essential for the production of food and fiber and for protection of public health and safety.

2. To protect the environment from environmentally harmful pesticides by prohibiting, regulating, or controlling uses of these pesticides.

3. To assure pest control workers of safe working conditions where pesticides are present.

4. To permit pest control by competent and responsible licensees and permittees under strict control of the California Department of Pesticide Regulation and county agricultural commissioners.

5. To assure the users that pesticides are properly labeled and are appropriate for the use designated by the label.

6. To encourage the development and implementation of pest management systems, stressing the application of biological and cultural pest control techniques, with selective pesticides when necessary, to achieve acceptable levels of control with the least possible harm to nontarget organisms and the environment.

must be a certified private applicator or a certified commercial applicator to use or supervise the use of restricted-use pesticides. In California, DPR is responsible for certifying pesticide applicators.

Other federal agencies, including the Department of Agriculture (USDA), the National Institute for Occupational Safety and Health (NIOSH), and the Fish and Wildlife Service (FWS), monitor and regulate some types of pesticide uses.

Throughout the United States, pesticide applicators and anyone handling pesticides must comply with federal laws. California has additional laws that are intended to make pesticide use safer for people and the environment under the special conditions existing here. Table 4-1 summarizes the responsibilities of government agencies in California's pesticide regulatory program. No U.S. state or locality can pass a law that allows pesticide uses or activities that the federal law prohibits. California and other states can, however, create pesticide laws that are more restrictive than federal laws. For example, California has a law that requires reporting of the use of all pesticides applied in many locations, including those treated by landscape maintenance gardeners. Federal laws do not require such reporting. This chapter summarizes the legal requirements you have to meet when you are applying pesticides or performing any pest control work for hire in turf, landscape, and interiorscapes.

California's pesticide and pest control laws are part of the California Food and Agricultural Code. The California legislature passes these laws in response to needs that arise within the state or in response to federal mandates. Regulations are the working rules set by state agencies to help them interpret and carry out these laws. Regulations pertaining to pest control and pesticide use are part of Title 3, California Code of Regulations. The California Department of Pesticide Regulation proposes these pesticide and pest control regulations when there is a need to interpret and carry out the intent of the laws. DPR makes proposals available for public comment through a "notice" process. DPR receives comments on a proposed regulation by mail and through public hearings. Once all comments have been considered, DPR submits final proposals to the Office of Administrative Law for approval. See Sidebar 9 for information on downloading copies of these laws and regulations as well as other resources from the internet.

Several other state agencies in California monitor and regulate pesticide use. These include the Department of Health Services, Air Resources Board, Department of Fish and Game, Department of Forestry, Occupational Safety and Health Administration, Waste Management Board, Department of Water Resources, Water Resources Control Board and Regional Water Control Boards, and Structural Pest Control Board. In addition, agricultural commissioners in each county develop pesticide use policies or conditions. These are specific to the needs of the individual county. Before they can become operative, county agricultural commissioner policies must be approved by the Director of DPR.

Laws and regulations change when new situations arise that are not covered by existing laws and regulations. For example, pesticides and pesticide application equipment are constantly being improved or changed. These changes often require that people use pesticides differently. Also, pest problems and pest management techniques change from time to time, and this, too, affects how people use pesticides. In addition, as regulators identify new health and environmental problems, they create or modify laws and regulations in order to deal with those problems.

TABLE 4-1.

Responsibilities of Government Agencies in California's Pesticide Regulatory Program.

REGULATORY FUNCTION	WHO ADMINISTERS IT*	WHAT IT DOES
Citing or prosecuting violators	EPA, DPR, CAC, SPCB, State Attorney General, local District Attorneys	Agricultural Commissioner may levy civil penalties with fines. DPR may request that Attorney General take civil action. Attorney General may file accusation. DPR may suspend or revoke applicator's certificate. CAC may suspend, revoke, or refuse permits and county registration.
Classifying pesticides	EPA, DPR	EPA classifies pesticides as restricted- or general-use in the United States. DPR may impose more stringent restrictions for California based on special conditions existing in the state.
Licensing commercial applicators, pest control advisoers, pest control businesses, dealers, and maintenance gardeners	DPR**	DPR issues licenses and administer tests to commercial applicator license applicants; also revokes, suspends, or refuses licenses upon violation of pesticide laws.
Monitoring pesticide residues on food and feed.	EPA, USDA, DPR, CAC***	Tests food and feed for pesticide residues; quarantines or destroys illegally contaminated commodities; brings cases of violation to county District Attorney or State Attorney General for prosecution.
Permitting	CAC, DPR	Issues, revokes, or refuses restricted-use (restricted material) pesticide permits (with use conditions) to growers, other private applicators, or certified applicators.
Pesticide disposal and storage	DPR, DHS, WRCB, WQCB, ARB, CAC	Regulates hazardous waste storage and disposal, pesticide container disposal sites, and water quality standards.
Pesticide illness investigation	DPR, CAC, DHS	Participates in pesticide illness investigations and in development of worker safety regulations; confirms compliance with worker training requirements.
Protecting wildlife	EPA, FWS, DFG, CAC, DPR	Investigates fish and wildlife losses; identifies and monitors endangered species; restricts pesticide use to protect endangered species and other wildlife.
Registering applicators and pest control advisers, certifying private applicators	CAC	Registers agricultural pest control businesses, aerial pest control operators, licensed pest control advisers, and maintenance gardeners. Through written examination, certifies private applicators. Provides applicators and advisers with information on local pesticide use conditions. Inspects pesticide use records and pest control recommendations to verify proper pesticide use.
Registering pesticides	EPA, DPR	Refuses or accepts registration; suspends, cancels, or reregisters pesticides.
Regulating pesticide use and worker safety	EPA, DPR, CAC, DHS	General authority to regulate pest control operations, including restrictions on the time, place, and manner of application; inspects training programs and records; various warning and enforcement powers.

*ARB = California Air Resources Board; CAC = County Agricultural Commissioner; DFG = California Department of Fish and Game; DHS = California Department of Health Services; DPR = California Department of Pesticide Regulation; EPA = US Environmental Protection Agency; FWS = US Fish and Wildlife Service; SPCB = Structural Pest Control Board, California Department of Consumer Affairs; USDA = US Department of Agriculture; WRCB = California Water Resources Control Board; WQCB = California Water Quality Control Board

**In California, Structural Pest Control Operators are licensed by the Structural Pest Control Board, California Department of Consumer Affairs. Vector Control Certificates are issued by the Department of Health Services.

***Thirteen other states and five federal agencies monitor various parts of the environment for pesticides and other substances.

REQUIREMENTS FOR LANDSCAPE MAINTENANCE PEST CONTROL QAC/QAL AND MAINTENANCE GARDENER LICENSING

People who perform pest control in landscape, turf areas, and interiorscapes are subject to certain California legal requirements. You must have a DPR Qualified Applicator Certificate (QAC) or a Qualified Applicator License (QAL) if you use or supervise the use of federally restricted-use pesticides or state-restricted materials see Sidebar 10. This applies to restricted pesticides used for any purpose or on any property. The only exception would be that provided by the definition of private applicator (see the Glossary for specialized definitions for these terms).

Any business that performs pest control for hire in landscapes, turf, or

SIDEBAR 9

Sources for Forms, Publications, and Information

ITEM	SOURCE
LAWS AND REGULATIONS	
☐ California Food and Agriculture Code	http://www.leginfo.ca.gov
☐ Department of Pesticide Regulation Laws and Regulations	http://www.cdpr.ca.gov
☐ Federal Insecticide, Fungicide, and Rodenticide Act (FIFRA)	http://www.gpoaccess.gov/multidb.html
PUBLICATIONS	
☐ *Laws and Regulations Study Guide*	http: www.cdpr.ca.gov or California Department of Pesticide Regulation 1001 I Street, P.O. Box 4015 Sacramento, CA 95812-4015
☐ *The Safe and Effective Use of Pesticides, Second Edition*	http://anrcatalog.ucdavis.edu or University of California, ANR Communication Services, 6701 San Pablo Avenue Oakland, CA 94608-1239 1-800-994-8849 or 510-642-2431
FORMS AND INFORMATION	
☐ Corporation registration information and *Certificate of Good Standing* form	http://www.ss.ca.gov/business/business.htm
☐ Pesticide use reporting forms	http://www.cdpr.ca.gov or County Agricultural Commissioner's Office
☐ *School Site Pesticide Use Reporting* form (PR-ENF-117) and information	http://www.cdpr.ca.gov
☐ Landscape Maintenance Pest Control and Maintenance Gardener licensing – forms and information	http://www.cdpr.ca.gov or County Agricultural Commissioner's Office
☐ Maintenance Gardener Pest Control licensing – forms and information, Business License – forms and information	http://www.cdpr.ca.gov or County Agricultural Commissioner's Office

interiorscapes must have an individual on its staff who has a QAL in the Landscape Maintenance Pest Control category. In addition, the business must obtain a Pest Control Business License from DPR. These requirements apply to the use of any type of pesticide, not just restricted materials.

Anyone who is in the business of landscape maintenance or works with interiorscapes and performs pest control incidental to that business must have a QAC or QAL in the Landscape Maintenance Pest Control category or the Maintenance Gardener Pest Control category. He or she must also obtain a Maintenance Gardener Pest Control Business License. These requirements apply to the use of any type of pesticide, not just restricted materials.

To qualify for aQAC or QAL in the Landscape Maintenance Pest Control category or Maintenance Gardener Pest Control category, you must

SIDEBAR 10

How to Obtain a Restricted-Material Permit

Once you have become qualified as a *certified applicator,* follow the steps below to obtain a restricted-use (restricted material) pesticide permit:

1. Telephone the agricultural commissioner's office in the county where the restricted material will be used. Make an appointment for issuance of a pesticide restricted-use (restricted material) permit.
2. Prepare a list of the use sites and restricted materials that will be used during the calendar year.
3. Prepare a list of pests to be controlled by the restricted materials.
4. Take all this information to the agricultural commissioner's office when applying for a restricted-use (restricted material) pesticide permit.

WHAT WILL HAPPEN

1. You will be asked to fill out an application and to verify that you hold a valid pesticide applicator's license or certificate (QAL or QAC).
2. The commissioner or a biologist will review some or all of the following information with you:
   a. pesticide use requirements
   b. methods of transporting, storing, and disposing of pesticides
   c. bee protection, if applicable
   d. environmental precautions
   e. equipment care and calibration
   f. posting requirements, if needed
   g. mixing and loading
   h. employee training and supervision
   i. protective clothing and equipment
   j. emergency medical requirements
   k. record keeping and reports
   l. county policies and requirements

AFTER THE PERMIT IS ISSUED

1. File a *Notice of Intent* with the county agriculture commissioner's office at least 24 hours before you make a restricted-use pesticide application. (You can make this notice in writing, in person, or by telephone. Some offices have a 24-hour telephone with an answering device to receive such messages.) Provide the following information:
   a. name and address of the permit holder and applicator
   b. permit number
   c. identification and location of the site or commodity to be treated
   d. approximate acres or other units of area
   e. application rate, dilution, and volume per acre
   f. pests to be controlled
   g. pesticide to be applied
   h. application method

i. date intended application is to begin
j. location and identity of areas that have been physically altered since the permit was issued and that may be adversely impacted by the application

2. Notify owners or managers of property being treated. Make sure that they are informed of the symptoms of pesticide poisoning and that they understand that they must not enter treated areas until the end of the restricted-entry interval.
3. If necessary, post warning signs indicating the area that is being treated (see Chapter 6).

AFTER THE PESTICIDE IS APPLIED

1. Dispose of pesticide containers according to instructions available from the county agricultural commissioner's office.
2. File *Pesticide Use Report* by the 10th of the month following the month of application.
3. Record pertinent information about the application and keep this information on file for two years.
4. Remove posted signs, if used, at the appropriate time.

successfully pass the corresponding examination administered by DPR. For a more thorough understanding of the laws and regulations and the basic principles examination, study these two guides:

- ***Laws and Regulations Study Guide.*** You can download the *Laws and Regulations Study Guide* from the DPR Web site at http://www.cdpr.ca.gov/. You can also purchase a printed copy of this publication from DPR.
- ***The Safe and Effective Use of Pesticides, Second Edition.*** This manual provides general information on pesticide use and handling. Many public libraries have this publication and you can purchase a copy of your own from the University of California, ANR Communication Services, 6701 San Pablo Avenue, Oakland, CA 94608-1239 (phone 1-800-994-8849 or 510-642-2431). You can order it online at http://anrcatalog.ucdavis.edu/.

The second exam deals specifically with landscape maintenance pest control. The manual that you are reading now, *Landscape Maintenance Pest Control*, is the recommended study guide for that examination.

The Landscape Maintenance Pest Control Category

A QAL or QAC in the Landscape Maintenance Pest Control category allows a person to use or supervise the use of general-use and restricted-use (restricted material) pesticides when performing pest control for hire to control pests in natural or planted landscape and turf areas, other outside areas around buildings, and interiorscapes. The QAL indicates that the person is qualified to supervise the operation of a licensed pest control business. Typical treatment areas included in this pest control category are residential yards (including home gardens and orchards), recreation areas, schoolyards, vacant lots, storage yards, greenbelts, and other similar sites that are landscaped or planted to ornamentals or turf. It also includes landscaped street medians and sidewalk areas and areas such as walkways and parking lots directly related to landscaped areas. In addition, this category covers landscaped areas in enclosed shopping malls and interiorscape plants. *This category does not include direct application to aquatic settings including running or standing water.*

Pest Control Business License

If you, whether as an individual or as a business, perform pest control for hire (including, but not limited to, advertising, soliciting, or operating as a landscape pest control business), you must possess a Pest Control Business License. This is true whether your office is the principal office or a branch location of the operation. Landscape pest control business types include

- exterior ground and/or aerial landscape pest control application businesses
- interior landscape pest control application businesses
- golf course management companies, when their service includes landscape pest control
- companies that perform landscape pest control on rights-of-way

Basic Licensing Requirements. You must have at least one person who is in a supervisory position in the business and holds a valid QAL with the appropriate pest control categories at each principal and branch location. You may obtain a Pest Control Business License by submitting an application, the appropriate fee, and the following supporting business information and documents:

- **Financial Responsibility Requirement.** This requirement can be met by providing DPR with copies of any of the following documents:
    - Certificate of Insurance, PR-PML-0173
    - Certificate of Insurance, PR-PML-052
    - *Acord Certificate of Liability Insurance* form from your insurance company stating that it meets and complies with Title 3, California Code of Regulations (3CCR) section 6524 requirements.
- **A Document that Verifies the Name and Type of Business.** You can provide copies of either of the following documents that apply to your license type:
    - *Fictitious Business Name Statement* from the County Clerk or County Recorder's office. This applies to any business operating under a fictitious name.
    - *Certificate of Good Standing* document from the California Secretary of State Office. This applies to any domestic or foreign corporation operating in California. The corporation must be registered with the California Secretary of State. See the Secretary of State Web site at http://www.ss.ca .gov/business/business.htm for registration information.
- **Workers' Compensation Insurance.** Each applicant who is an employer as defined in Section 3300 of the Labor Code is required to carry workers' compensation insurance.

Before you conduct any work for hire, you must register your Pest Control Business License with the county agricultural commissioner's office in each county in which your business intends to perform pest control. Most counties require a fee for registration. Registration must be renewed annually and covers one calendar year.

Pest Control Business Licenses may be suspended or revoked and the holder of this license may be subject to fines should the business fail to

- retain pest control application notification records for two years
- retain records of pesticide use for two years
- submit pesticide use reports (if needed) to the county agricultural commissioner's office in the county

where the application was made
- have valid permits for restricted materials used
- retain agricultural pest control adviser written recommendations (required when making applications for public agencies) for one year

Once you have obtained the Pest Control Business License, you are required to
- retain records of pesticide use for two years
- regularly submit pesticide use reports to the county agricultural commissioner
- have valid permits for restricted materials used
- pay a yearly fee

For additional information about the Pest Control Business License and to obtain forms to apply for this license, contact your county agricultural commissioner's office or visit the DPR Web site at http://www.cdpr.ca.gov/.

The Maintenance Gardener Pest Control Category

Maintenance gardeners who apply any general-use or restricted-use (restricted material) pesticide for pest control incidental to their maintenance gardening business must obtain a Maintenance Gardener Pest Control Business License. To qualify for this business license, the maintenance gardener business needs a person who possesses a valid QAC or QAL in the Maintenance Gardener Pest Control category or the Landscape Maintenance Pest Control category. This business license limits the gardener to incidental pest control in landscape and turf plantings indoors and outdoors, such as on private residences, condominiums, apartments, businesses, commercial parks, and plantings surrounding structures. DPR considers commercial parks and plantings surrounding structures to mean the following:

- A *commercial park* is a tract of land where several commercial establishments (e.g., stores, offices, restaurants, warehouses, factories) are located. They include business parks, industrial parks, office parks, and shopping centers or malls.
- *Plantings surrounding structures* that are ornamental and turf plantings that are incidental to and placed near or close to structures such as buildings, brick walls, fountains, fences, or statues.

Maintenance Gardener Pest Control Business License

You must meet specific requirements in order to obtain a Maintenance Gardener Pest Control Business License. These include
- having at least one person in a supervisory position who holds a valid QAC or QAL with the Maintenance Gardener Pest Control category or the Landscape Maintenance Pest Control category at each business location
- fulfilling the financial responsibility requirements outlined in the California Food and Agricultural Code and the California Code of Regulations (see Pest Control Business License information above) or submitting a Liability Certificate Statement, PR-PML-170, indicating you are financially able to respond to damages using your own assets
- obtaining a Fictitious Business Name Statement from the County Clerk's Office (see Pest Control Business License information above)
- (for companies that are corporations) obtaining a Certificate of Good Standing from the California Secretary of State (see Pest Control Business License information above)
- paying a yearly license fee

Once you have obtained your Maintenance Gardener Pest Control Business License, you are required to

- register with the county agricultural commissioner in each county where you perform work
- retain records of pesticide use for two years
- regularly submit pesticide use reports to the county agricultural commissioner
- have valid permits for restricted materials used

For additional information about the Maintenance Gardener Pest Control Business License and to obtain application forms for this license, contact your county agricultural commissioner's office or visit the DPR Web site at http://www.cdpr.ca.gov/.

Exemption from the Maintenance Gardener Pest Control or Pest Control Business License Requirements

People who perform pest control incidental to new construction are not required to obtain the Maintenance Gardener Pest Control Business License or the Pest Control Business License. Examples of this type of work include construction work such as

- paving parking lots or driveways
- establishing new landscape
- building homes or other structures

ENFORCEMENT AND COMPLIANCE ACTIONS

The Department of Pesticide Regulation and county agricultural commissioners have the authority to enforce pesticide use laws and regulations, including taking enforcement and compliance actions against violators at the county level. DPR has the authority at the state level to take further action against violators. Compliance actions may consist of

- a cease and desist order
- a violation notice
- a warning letter
- a compliance interview

Enforcement actions may consist of

- administrative penalties consisting of fines of up to $5,000 per violation
- disciplinary action (suspension or revocation) against a county registration to perform work or against a state-issued certificate or license
- civil court action resulting in fines from $1,000 to $25,000 per violation

In addition to actions taken by DPR and county agricultural commissioners, the pesticide regulations authorize city and county district attorneys to take criminal court actions that may result in fines from $500 to $50,000 per violation and imprisonment for between 10 days and one year.

PESTICIDE LABELS

A pesticide's label is the legal documents that you are required to follow when using or handling that pesticide. The pesticide label provides you with the information necessary for safe and proper handling, storage, and disposal of the pesticide. You are required by law to have the complete label at the application site for each pesticide being used.

Regulations set the format for pesticide labels and prescribe what information they must contain. Some pesticide packages are too small to have all of this information printed on them, so in such cases U.S. EPA requires that manufacturers attach a *supplemental label* (Figure 4-1). For plastic containers, the supplemental label is put in a plastic pouch attached to the side of the container. Paper packages usually have a supplemental label inserted under the bottom flap.

The Safe and Effective Use of Pesticides, Second Edition contains a sample pesticide label and describes the information you will find on various parts

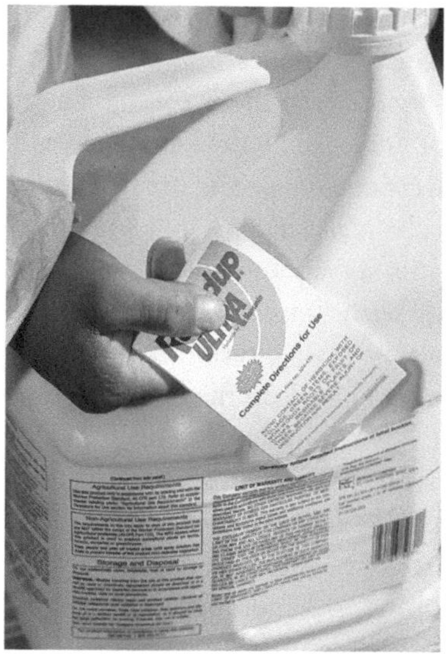

FIGURE 4-1.

Supplemental labels often are attached to pesticide packages. Before you purchase a pesticide, make sure that you have all of the labeling associated with that material.

of the label. Become familiar with all of the information that is included on a pesticide label.

When to Read the Pesticide Label

Pesticide labels contain important information you need to understand before you use the product. At the very least, read the pesticide label (Figure 4-2) at these times:

- *Before you buy the pesticide.* Make sure the pesticide is registered for your intended use. Confirm that there are no restrictions or other conditions that prohibit you from using the pesticide at the application site. Be certain that your intended use is suitable under current weather conditions. Also, make sure that the pesticide controls the current life stage of your pest. Find out what protective equipment and special application equipment you will need.
- *Before you mix and apply the pesticide.* Read the label to learn the mixing and application directions. Find out what precautions to take to prevent pesticide exposure to people and nontarget organisms. Learn what first aid and medical treatments are necessary should an accident occur.
- *When storing pesticides.* Read the label directions dealing with how to store the pesticide properly. Understand the special precautions to prevent fire hazards.
- *Before disposing of unused pesticide and empty containers.* Learn how to prevent environmental contamination and hazards to people. (Before disposal, check with the agricultural commissioner in your county for local restrictions and requirements.)

Deviations from Label Directions

You must follow the label directions when using any pesticide. California regulations do, however, allow six specific label deviation situations. These situations generally involve safer or less disruptive uses of the pesticide. Following are the only label deviations allowed by state regulations. These

FIGURE 4-2.

The pesticide label is a complex legal document that you must read and understand before you make a pesticide application. It must be present at each use site during the application. Make all pesticide applications in strict accordance with label instructions.

exceptions may change at any time and they may not apply in certain instances. Always check with DPR or your county agricultural commissioner before using a pesticide in any manner that deviates from label directions.

- *Decreasing the rate of pesticide applied.* It is allowable to apply a lower application rate than what the label recommends. One of the reasons for doing this would be to protect beneficial insects that might contribute to control of the pest. Under no circumstances can you legally increase the application rate of the pesticide you apply beyond the maximum label rates.
- *Decreasing the concentration of the mixture applied.* Label directions usually state the concentration of the mixture as the volume of water or other diluent to use when preparing a spray mixture. It is always possible to use more water or other diluent than this, but if you use too much diluent you may end up with excessive dilution and runoff. If this happens, too little pesticide may actually get to the target pest. In most cases, you should use only as much water or other diluent as necessary to obtain thorough coverage, but make sure not to use less than what the label states.
- *Applying at a frequency less than specified.* Label directions often prescribe how often to apply a pesticide. Manufacturers recommend that you re-apply a pesticide at specified intervals in order to maintain adequate control of the pest. It is permissible under the law to make applications less frequently than the label recommends. If your monitoring confirms that less-frequent applications provide adequate pest control, there is no need to treat more often. Additional applications sometimes increase other pest problems by disrupting beneficials. Besides, it is expensive to use unneeded pesticides.
- *Using the pesticide to control a target pest not listed on the label, so long as the commodity or site is listed on the label and use against an unnamed pest is not expressly prohibited.* You may wish to use a pesticide on turf, landscape, or interiorscape plants that are listed on the label, though the label does not list the target pest. In this situation, so long as the label does not forbid the use of this pesticide against the pest on the turf, landscape, or interiorscape plants, you may use it. Follow all other label directions.
- *Using any method of application not prohibited, provided that other label directions are followed.* Most labels do not specify exactly how to apply the pesticide. Should this be the case, it is permissible to use any practical method, so long as it is not prohibited, but you must still make sure that the method you choose allows you to follow all other label directions.
- *Mixing with another pesticide or fertilizer, unless prohibited.* You may want to combine one pesticide with one or more others or with fertilizers. This practice saves time and reduces application costs. Unless specifically prohibited by the directions on any of the labels, it is permissible to apply pesticides in combination.

NOTICE OF APPLICATION

You must give notice to the property owner or operator before you perform any pest control. This notice is especially important if you are performing the pest control for hire and other people unknown to you have access to the property you will be treating. When the property is used for commercial or research production of agricultural commodities, you are required to give considerable information to the property manager. If the pest control is in landscapes, interiorscapes, or on turf that fall within the scope of the Landscape Maintenance Pest Control category, however, this notice is limited to

- the date of the scheduled application
- the brand name or common chemical name of the pesticide to be applied
- the precautions people entering the property must observe (from the printed label or laws and regulations)

PESTICIDE USE RECORDS

Under certain circumstances, regulations require that you maintain records of your pesticide use if you apply pesticides to landscapes, interiorscapes, or turf. These circumstances include

- making pesticide applications for hire
- using any restricted-use pesticide (restricted material) or any agricultural-use pesticide

Keep these records for two years and be ready to make them available to the Director of DPR or the county agricultural commissioner upon request. Include the following information (Figure 4-3) in your pesticide use records:

- date of application
- name of the operator of the property treated
- location of property treated
- crop commodity or site treated
- total acreage or units treated at the site
- pesticide used, including the U.S. EPA or state registration number that is on the pesticide label
- amount used

School Sites. If you apply pesticides for hire at a school site (kindergarten

STATE OF CALIFORNIA
MONTHLY SUMMARY PESTICIDE USE REPORT
PR-ENF-060 (REV. 4/92)
INSTRUCTIONS FOR COMPLETING THIS FORM ARE INDICATED BELOW

DEPARTMENT OF PESTICIDE REGULATION
PESTICIDE ENFORCEMENT BRANCH

OPERATOR (FIRM NAME)		ADDRESS	CITY	ZIP CODE	PHONE NUMBER
OPERATOR ID/PERMIT NUMBER	LICENSE NUMBER	COUNTY WHERE APPLIED	COUNTY NUMBER	MONTH/YEAR OF USE	TOTAL NUMBER OF APPLICATIONS

1. Complete Columns A, B, C, and D for All Users
2. Complete Column E by Using one of the Following Codes
Code 10 - Structural Pest Control..................includes any pest control work performed within or on buildings and other structures.
Code 30 - Landscape Maintenance Pest Control..........includes any pest control work performed on landscape plantings around residences, or other buildings, golf courses, parks, cemeteries, etc.
Code 40 - Right-of-Way Pest Control..................includes any pest control work performed along roadsides, power lines, median strips, ditch bands and similar sites.
Code 50 - Public Health Pest Control..................includes any pest control work performed by or under contract with State or local public health or vector control agencies.
Code 80 - Vertebrate Pest Control..................includes any pest vertebrate pest control work performed by public agencies or work under the supervision of the State or county agricultural commissioner.
Code 91 - Commodity Fumigation (Nonfood/Nonfeed).includes fumigation of nonfood/nonfeed commodities such as pallets, dunnage, furniture, burlap bags, etc.
Code 100 - Regulatory Pest Control..................includes any pest control work performed by public employees or contractors in the control of regulated pests.
3. Complete Columns F and G, if Use Does not Fit one of the Above Codes

A	B	C	D	E	F	G
MANUFACTURER AND NAME OF PRODUCT APPLIED	EPA/CALIFORNIA REGISTRATION NUMBER FROM LABEL (INCLUDE ALPHA CODE)	TOTAL PRODUCT USED (Check One Unit of Measure)	NUMBER OF APPLICATIONS	CODE	COMMODITY OR SITE TREATED	ACRES/UNITS TREATED
	– – –	☐ LB ☐ OZ ☐ PT ☐ QT ☐ GA				
	– – –	☐ LB ☐ OZ ☐ PT ☐ QT ☐ GA				
	– – –	☐ LB ☐ OZ ☐ PT ☐ QT ☐ GA				
	– – –	☐ LB ☐ OZ ☐ PT ☐ QT ☐ GA				
	– – –	☐ LB ☐ OZ ☐ PT ☐ QT ☐ GA				
	– – –	☐ LB ☐ OZ ☐ PT ☐ QT ☐ GA				
	– – –	☐ LB ☐ OZ ☐ PT ☐ QT ☐ GA				
	– – –	☐ LB ☐ OZ ☐ PT ☐ QT ☐ GA				

REPORT PREPARED BY ______________________ DATE ______________

Reports must be submitted to the county agricultural commissioner by the 10th of the month following the month in which the work was performed.

FIGURE 4-3.

Any pesticide applied either for hire or incidental to landscape maintenance or maintenance gardener activities, must be reported in writing. Submit reports monthly to the county agricultural commissioner for the county of application on a form such as this. See Sidebar 9 for information on downloading the form.

through 12th grade), you must also include the following information in your records:

- time application was completed
- name and address of school site
- application location at the school site (such as athletic fields, landscaped areas, etc.)

OPERATOR IDENTIFICATION NUMBER

If you are a property operator who is involved in the production of agricultural commodities or who applies pesticides in certain other areas, you must follow the procedures described below before you buy or use pesticides. If you perform pest control for hire, however, you are exempt from these requirements. (In this context, performing pest control "for hire" means that you are a self-employed pest control operator or an employee of a firm that contracts to apply pesticides for others and that you or your employer do not own, manage, or operate the site where you apply the pesticides.)

Generally, there are two circumstances when property operators must obtain an Operator Identification Number before they buy or use pesticides:

1. when using pesticides for the production of an agricultural commodity
2. when using a pesticide for an agricultural use, but not producing an agricultural commodity (e.g., cemeteries, golf courses, parks, and rights-of-way)

Pesticide use for certain other non-agricultural sites requires an Operator Identification Number. If someone hires you to perform pest control, whether as part of your pest control business or incidentally as part of your landscape maintenance or maintenance gardener business, you are not required to obtain an Operator Identification Number.

Obtain an Operator Identification Number from the agricultural commissioner's office of each county in which you perform pest control. If you are the operator or an employee of the operator of the property, you are not required to obtain an Operator Identification Number when a person or company that performs pest control for hire purchases and applies these pesticides.

Following are the pesticides for which you must have an Operator Identification Number if you apply them on turf, landscape, interiorscape, and other nonagricultural sites that you operate.

- any pesticide designated for agricultural use (e.g., cemeteries, golf courses, parks, and rights-of-way), as defined in the Food and Agricultural Code Section 11408; pesticides designated for use only on livestock are exempted
- any pesticide designated by the DPR as Restricted (listed in Section 6400 of the California Code of Regulations)
- any pesticide designated for industrial use as a postharvest commodity treatment
- any pesticide listed on the Groundwater Protection List found in Section 6800(b) of the California Code of Regulations and used for any outdoor institutional or outdoor industrial use

Monthly Summary Pesticide Use Reports

Monthly summary pesticide use reports are required if you

- apply any pesticide designated for agricultural use, as defined in the Food and Agricultural Code Section 11408 (e.g., cemeteries, golf courses, parks, and rights-of-way); pesticides designated for use only on livestock are exempted
- apply pesticides for hire or as an incidental element of landscape maintenance or maintenance gardener activities

- use a pesticide for industrial post-harvest commodity treatment
- use any pesticide listed on the Groundwater Protection List found in Section 6800(b) of the California Code of Regulations, for any outdoor institutional or outdoor industrial purpose

If you meet the criteria listed above and apply pesticides to turf, landscapes, interiorscapes, or other nonagricultural sites, you must report a summary of your monthly use of pesticides to the agricultural commissioner of the county in which you performed the work. Deliver this report by mail or in person to the county agricultural commissioner's office. Submit the report by the 10th day of the month following the month in which you applied the pesticides. Include the following information on this report:

- name and address of the person (or business) who applied the pesticides
- county where the pest control was performed
- month and year of pesticide use
- crop, commodity, or site treated, except when using a designated use code (as specified on the Monthly Summary Pesticide Use Report form)
- the pesticide names, including the U.S. EPA or state registration numbers on the pesticide labels
- amount of each pesticide used
- number of applications made with each pesticide and total number of applications made during the month
- total acres or units treated with each pesticide, except when using a designated use code (as specified on the Summary Pesticide Use Report form)

You can get reporting forms from your county agricultural commissioner's office or download them from DPR's Web site at http://www.cdpr.ca.gov.

Yearly Pesticide Use Reports for School Sites

In addition to the monthly summary report of pesticide usage, if you apply pesticides for hire at a school site for any grade Kindergarten through 12th grade (K-12) and are required to maintain pesticide use records, you must submit an annual report of pesticide use to the DPR Director. File your report for each year by January 30 of the succeeding year. Use the School Site Pesticide Use Reporting form (PR-ENF-117) available from DPR. You can download this form and get more information about this requirement from the DPR Web site at http://www.cdpr.ca.gov/. You must include the following information on this yearly report:

- name and address of the person or business applying the pesticides
- county where the pest control was performed
- date and time of pesticide use
- school site identification number or the name and address of school site
- location of pesticide application
- name of pesticide, including the U.S. EPA or state registration number from the pesticide label
- amount of pesticide used

RESTRICTED-USE PESTICIDES (RESTRICTED MATERIALS)

Federal and state laws establish hazard criteria for classifying pesticides as restricted-use pesticides (the federal classification) and as restricted materials (the California classification). These are pesticides that, even when properly used according to the registered label, may be expected to cause an unreasonable adverse effect on the applicator and on other workers. They may also cause an unreasonable adverse effect on the environment, including wildlife, bees, and domestic animals. The Federal Insecticide, Fungicide, and Rodenticide Act (FIFRA) defines "unreasonable effects on the environment" as ". . . any unreasonable risk to man or the environment, taking into account the economic, social, and environmental costs and benefits of the use of any pesticide . . . ." To mitigate these adverse effects, these pesticides may be covered by California regulatory use directions and restrictions that exceed the use directions and restrictions found on the pesticide label. Counties may also place local restrictions on particular pesticides. Information on federal, state, and county use restrictions are available from the county agricultural commissioner's office in the county where you make an application. Criteria for restricting the use of certain pesticides in California include the following:

- danger to public health
- hazard to applicators and farm workers
- hazard to domestic animals, honey bees, and crops from direct application or drift
- hazard to the environment from drift onto streams, lakes, or wildlife estuaries
- hazards related to persistent residues in the soil resulting in contamination of the air, waterways, estuaries, or lakes, causing damage to fish, wild birds, and other wildlife
- hazards to subsequent crops due to persistent soil residues
- a federal restricted pesticide

Restricted Material Permit

Generally, and with few exemptions, a property operator must possess a valid restricted material permit before he or she can purchase, possess, or use restricted materials. The county agricultural commissioner is responsible for issuing restricted material permits to the property operator or to an authorized representative. Restricted material permits may be issued for agricultural or nonagricultural uses. To be issued a permit, the applicant must be a certified private applicator or certified commercial applicator. The permit may be issued to the property operator's authorized representative instead, as long as that person meets the certified private or commercial applicator requirement.

No restricted material permit is required for

- a registrant, a pest control dealer possessing restricted materials under his or her license, or a commercial carrier when transporting restricted materials
- a certified private or commercial applicator possessing or using only federally restricted materials
- a person possessing or using only antifouling paints or paints containing tributyltin
- a person possessing or using restricted materials, but only for research or experimental work conducted by college or university research personnel
- a person possessing or using restricted materials under a valid research authorization issued by DPR

Agricultural restricted material permits are site and time specific. They require detailed information, including, but not limited to

- location of the use site or commodity to be treated
- pests to be controlled
- identification of non-target sites around and near the area to be treated that may be adversely impacted by the proposed treatment
- date and time of proposed treatments
- method of applications
- the restricted materials to be applied to each site as well as application and dilution rates
- the name of the pest control business, if any, and name, business address, and license or certificate number, with expiration date, of

the certified private or certified commercial applicator responsible for supervising the possession or use of the restricted materials

Nonagricultural restricted material permits require less information than is required for agricultural restricted material permits. The required information includes

- name and business address of the operator of the property or the pest control business
- restricted materials to be used
- pests to be controlled
- method of application
- criteria for determining the need for the pesticide application
- the name, address, and license or certificate number, with expiration date, of the certified applicator responsible for supervising the possession or use of the restricted materials

Notice of Intent

If certain information is not included in the permit, the permit holder must give notice to the county agricultural commissioner at least 24 hours prior to the intended application. The notice must contain information including, but not limited to,

- name and address of the permit holder and applicator
- permit number
- identification and location of the site or commodity to be treated
- approximate acres or other units
- application rate, dilution, and volume per acre
- pests to be controlled
- pesticide to be applied
- application method
- date intended application is to begin
- location and identity of areas that have changed since the permit was issued and that may be adversely impacted by the application

The notice of intent may be submitted to the county agricultural commissioner by the operator of the property to be treated. Alternatively, it may be submitted by the operator's authorized representative or by the licensed pest control operator who is to apply the pesticide.

APPLICATOR CERTIFICATION

California has specific requirements that must be met by anyone wishing to become a certified pesticide applicator. In general, you must be a certified private applicator or certified commercial applicator in order to purchase, store, use, or supervise the use of restricted materials. The two commercial applicator certifications administered by DPR are the Qualified Applicator Certificate (QAC) and the Qualified Applicator License (QAL). Either certification allows you to use or supervise the use of restricted materials in settings not covered by the definition of a Certified Private Applicator.

You must possess a valid QAL or QAC in the Landscape Maintenance Pest Control category before you can perform

- pest control for hire as part of your pest control business
- incidental pest control for hire as part of your maintenance gardener pest control business

This requirement applies regardless of whether you use general-use pesticides or restricted-use pesticides (restricted materials).

In certain agricultural-use settings, a valid QAC or QAL in the Landscape Maintenance Pest Control category will allow you to use or supervise the use of restricted materials in landscape maintenance pest control as the employee of an employer that does not possess a private applicator certificate.

In some cases, you may also need a valid restricted material permit. See *Restricted Material Permit* later in this chapter.

To obtain a QAL or QAC, you must pass two examinations. The passing score for each exam is 70%. One examination tests your knowledge of pesticide use laws and regulations and the basic principles of safe and effective pesticide use. The second tests your knowledge of pest control in landscape maintenance pest control settings.

The California Department of Pesticide Regulation administers these examinations on a regular basis. Obtain applications and information for these examinations from your county agricultural commissioner's office or from the DPR Web site, http://www.cdpr.ca.gov

Renewing Your QAC or QAL. You may renew your QAC or QAL in either of two ways:

- Submit the appropriate application fee and examination fees and then pass the required examinations.

OR

- Earn sufficient continuing education hours during the valid period of your QAC or QAL and pay the required renewal fees.

You must earn the continuing education hours from DPR-approved training sessions. To avoid paying a late renewal fee, make sure to submit proof of session attendance and your renewal fee before the end of your current certificate's valid period.

Generally, if you possess a QAL or QAC in the Landscape Maintenance Pest Control category, you must earn 20 hours of continuing education every two years, with at least four hours covering pesticide use laws and regulations. If your QAC or QAL is for performing incidental pest control in the Landscape Maintenance Gardener Pest Control subcategory for maintenance gardeners, you need earn only 8 hours of continuing education every two years, with at least two of these hours covering laws and regulations.

Continuing Education

Continuing education is an important way for you and your employees to keep current on new information and review old knowledge. This includes issues of pesticide use, pesticide safety, and pest management. Keeping current is necessary because

- lawmakers change or revise pesticide laws and regulations from time to time
- pesticide chemicals and their formulations change often and new chemicals are introduced each year
- pesticide use practices change with increased knowledge of pests and their habits
- pesticide application techniques may change as new application equipment technologies and formulation types are developed
- researchers modify pest management techniques as they discover new information about pests
- new pests may arise, pests previously of lesser concern may become primary pests, and other pests may show resistance to pesticides that were previously more effective for their control

Without continuing education, you can quickly lose touch with current pest management information, modern pest control practices, and pesticide application techniques.

A good way to keep yourself current is to attend meetings of a professional organization affiliated with your work. Most pesticide-related occupations have one or more professional organizations such as the Pesticide Applicators Professional Association (PAPA) (see Sidebar 11). Several of these associations publish newsletters or sponsor workshops covering current and new pest management and pesticide use information. In addition, county UC Cooperative Extension horticulture and farm advisors sponsor workshops and seminars on aspects of pest control, pest management, and pesticide use. County agricultural commissioners' offices regularly conduct meetings to update people on pesticide laws and regulations. Special courses are also available through local community colleges, state colleges and universities, and the University of California.

Continuing education courses must be approved by the DPR if you are to receive QAL or QAC credit for attending. DPR assigns identification numbers to approved courses. When renewing your license or certificate, list each course name and identification number as well as the number of hours attended on the renewal application.

SIDEBAR 11

Contacts for Continuing Education Courses

Pesticide Applicators Professional Association (PAPA)

P.O. Box 80095
Salinas, CA 93912-0095
tel: (831) 442-3536
fax: (831) 442-2351
http://www.papaseminars.com

California Association of Pest Control Advisers (CAPCA)

1143 N. Market Boulevard, Suite 7
Sacramento, CA 95834
tel: (916) 928-1625
fax: (916) 928-0705
http://www.capca.com/start1.asp

California Department of Pesticide Regulation (DPR)

dpr California Environmental Protection Agency
Department of Pesticide Regulation

http://www.cdpr.ca.gov/docs/license/classes.htm

CHAPTER 4: REVIEW QUESTIONS

1. The federal laws regulating the manufacture, sale, transportation, and use of pesticides are administered by

☐ a. USDA
☐ b. NIOSH
☐ c. DPR
☐ d. U.S. EPA

2. Regulations pertaining to pest control and pesticide use in California are part of

☐ a. the Federal Insecticide, Fungicide, and Rodenticide Act (FIFRA)
☐ b. EPA Worker Protection Standards
☐ c. Title 3 of the California Code of Regulations
☐ d. Fish and Wildlife Service Regulations

3. The laws and regulations controlling pesticide use in California

☐ a. are optional as long as federal regulations are followed
☐ b. allow some pesticide handling practices prohibited at the federal level
☐ c. may be more restrictive than federal laws and regulations
☐ d. are exactly the same as the federal laws and regulations

4. What does the "Statement of Practical Treatment" on the pesticide label tell you?

☐ a. first aid instructions
☐ b. PPE requirements
☐ c. application instructions
☐ d. mixing and loading instructions

5. Which of the following deviations from label directions is legal in California?

☐ a. applying less than the label rate
☐ b. applying more than the label rate
☐ c. treating a site not on the label
☐ d. making more frequent applications than the label allows

6. How long must you keep pesticide use records?

☐ a. 6 months
☐ b. 1 year
☐ c. 2 years
☐ d. 5 years

7. Regulations are

☐ a. laws that regulate how pesticides can be used
☐ b. pesticide use policies or conditions
☐ c. ordinances governing pesticide use
☐ d. working rules needed to carry out laws

8. A QAC in the Landscape Maintenance Pest Control Category allows a person to

☐ a. supervise the operation of a licensed pest control business
☐ b. use or supervise the use of general or restricted-use pesticides in turf, landscape, and interiorscape settings
☐ c. make pesticide applications for hire in turf, landscape, and interiorscape settings
☐ d. make pesticide applications for hire on rights-of-way

9. Once you obtain a Pest Control Business License, you must retain records of pesticide use for

☐ a. 1 year
☐ b. 2 years
☐ c. 3 years
☐ d. 5 years

10. The Maintenance Gardener Pest Control category allows individuals to

☐ a. own a pest control business
☐ b. use general-use and restricted-use pesticides incidental to their landscape maintenance business
☐ c. become a pesticide dealer
☐ d. supervise the operation of a pest control business

11. A Pest Control Business License is not required if you

☐ a. make pesticide applications for hire
☐ b. perform pest control incidental to new construction
☐ c. contract to control landscape pests on school grounds
☐ d. contract to control landscape pests in cemeteries

12. When you perform pest control in turf, landscape, or interiorscape settings, you must

☐ a. give prior notice to the property owner or operator before applying any pesticide
☐ b. notify the property owner or operator after completing the application
☐ c. give prior notice to the property owner or operator only before applying any restricted-use pesticide
☐ d. give prior notice to anyone using the area before applying any pesticide

13. The minimum number of hours of continuing education required for a person with a QAC in the Landscape Maintenance Pest Control category (except those in the subcategory of maintenance gardener)

☐ a. 8 hours every year
☐ b. 8 hours every 2 years
☐ c. 20 hours every year
☐ d. 20 hours every 2 years

5 Hazards Associated with Pesticide Use

Many pesticide chemicals are hazardous and may be capable of injuring people, animals, and the environment. You must be able to recognize actual or potential hazards associated with pesticide applications in turf, landscape, and interiorscape situations. Pesticide handlers and other landscape maintenance workers are at the highest risk for illness and injury, but the public can also be exposed to pesticides during applications and to their residues afterward. Urban uses of pesticides have also led to the exposure of domestic animals and wildlife and have seriously contaminated ground and surface waters, soil, and ornamental and natural plantings. Human exposure and environmental contamination have prompted regulatory restrictions for pesticide uses in urban settings.

This chapter focuses on the following specific hazards associated with using pesticides in landscape, turf, and interiorscape situations:

- human exposure
- injury to pets (including pond fish)
- contamination of surface water
- contamination of groundwater
- injury to plants
- damage to objects in the treated area
- impact on wildlife and nontarget organisms

In addition to understanding the hazards listed above, you should review the general hazards associated with pesticide use as set forth in *The Safe and Effective Use of Pesticides, Second Edition.*

HUMAN EXPOSURE

Landscaped areas, turf, and buildings that have interiorscapes are sensitive areas because they are places that people frequently visit and may be densely populated. By law, you must protect the public as well as pesticide handlers and other employees from any type of exposure during and after any pesticide application.

Pesticides, like other poisonous chemicals, injure people by interfering with certain body functions. The type and extent of injury depend on the pesticide's toxicity and the amount (dose) entering the body's tissues. Some pesticides are very toxic and a low dose can cause an injury that appears either immediately or sometime later. A small exposure to one of these pesticides might cause a person serious health problems. Other pesticides are so slightly toxic that a person would have to receive a larger dose before he or she would experience any effect.
Because potential hazards exist with most pesticides and each individual may react differently to various levels of exposure, anyone handling or applying them must avoid exposure. Additionally, if the label indicates a period of time that must elapse after application before people can safely enter the treated area, you must protect any people who would otherwise pass through, use, or in any way come into contact with the treated area until that time has passed.

How People Get Exposed to Pesticides

Pesticide exposure occurs in several ways, such as through

- pesticide handling activities
- drift
- contact with treated surfaces
- accidents

Pesticide Handling Activities

People who handle pesticides are most at risk for pesticide exposure because they work closely and frequently with these materials. The most serious injuries usually occur from exposure during mixing and application. Every year, pesticide handler injuries make up the largest portion of the pesticide-related illnesses and injuries reported in California.

People who maintain or repair application equipment may contact pesticide residues that remain on the equipment. Frequent cleaning of the application equipment lowers risks to maintenance workers and operators, since cleaning prevents these residues from building up.

The wearing of proper work clothing and use of other protective equipment during mixing and application greatly reduce exposure risks. Always read and follow all of the instructions and precautions on the pesticide label of the product you are using. Refer to the next chapter for more information on personal protective equipment (PPE).

Drift

Drift is that portion of the applied pesticide droplets or dust that blows away from the intended application site while you are applying pesticides to target plants or turf. If drift occurs during an application, the droplets or dust may contact nearby people, animals, or objects. Drift can be caused by any excessive air movement, be it wind during applications made outside or air currents created by a ventilation system when applications are made to interiorscapes.

Another type of off-site movement that is possible in various settings occurs when outdoor applications are made near the intake for air conditioning, heating, or ventilation systems. Take extreme care to keep pesticides from entering any ventilation intake and so prevent their spread to other areas that people may occupy or pass through.

To protect people from drift, keep them away from the area you are treating. Use "caution" tape or place signs around the treatment site to help keep unauthorized people out. You may wish to establish a buffer strip—an area that provides adequate distance between the treatment site and the area people can safely occupy without coming in contact with any pesticide that may move off site. See Chapter 6 for posting requirements to follow when you apply pesticides in landscape, turf, and interiorscape settings.

Factors that influence drift include

- air movement (natural or artificial)
- application droplet or particle size
- spray pressure (for liquids)
- spray nozzles—number, placement, and configuration (for liquids)
- distance that dust or droplets must travel

Air Movement. Air movement is by far the most frequent and important cause of drift (Figure 5-1). The higher the speed, the more certain it is that drift will occur. Air movement in excess of 10 miles per hour is certain to cause drift. To prevent air from carrying pesticide droplets or dust off-site, never make pesticide applications when the air speed exceeds 10 mph.

Application Droplet or Particle Size. The size of applied droplets or particles affects the ability of spray or dust to drift (Figure 5-2). Smaller droplets or dust particles are lighter and more likely to drift. With liquid formulations, the spray nozzle orifice size controls the size of spray droplets. Smaller

FIGURE 5-1.

Air movement is the most important cause of pesticide drift. The higher the wind speed, the more certain drift will occur.

FIGURE 5-2.

Droplet or dust particle size affects the ability of spray droplets or dust to drift. Smaller droplets or dust particles are lighter and more likely to drift.

FIGURE 5-3.

Spray pressure is a contributing factor to drift of liquid sprays. As spray pressure increases, droplet size gets smaller.

orifices produce smaller droplets. To reduce the amount of droplet drift, always use the largest nozzle size that is practical for your spray application.

Spray Pressure. Spray pressure also contributes to the drift of liquid sprays because it influences droplet size (Figure 5-3). As spray pressure increases, droplets get smaller. It is best to use low-pressure nozzles on your spraying equipment and keep the spray pressure within the range recommended by the nozzle manufacturer.

Spray Nozzles. An increase in the number of nozzles also increases the possibility that more spray material will drift off site (Figure 5-4). The configuration of the nozzles also influences drift potential.

Distance that Dust or Droplets Must Travel. When you are applying liquids or dusts, the distance between the application equipment and the target plants has an influence on drift potential (Figure 5-5). As the distance a droplet or dust particle must travel increases, the chances of drift increases because air currents have more time and so more opportunity to move the droplets or particles. In addition, liquid droplets begin to evaporate once they leave the spray nozzle, so the further they travel, the smaller and lighter they become.

Contacting Treated Surfaces

Once you make an application, a residue of the pesticide remains on the treated surfaces for a while. Anyone who touches the treated plants or other surfaces is likely to pick up some of this residue on his or her hands or clothing. Someone who touches plants treated with dusts may stir up dust particles and possibly inhale them. To reduce the possibility of such exposure, keep people out of the area after the application for as long as the label indicates (Figure 5-6). Notify anyone using the area that they should not contact any treated plants or other items or surfaces. In interiorscapes, avoid getting residues onto furniture.

Accidents

The most harmful pesticide exposure risks occur during accidents that involve spills, splashes, rupture of

FIGURE 5-4.

An increase in the number of nozzles also increases the potential volume of spray material subject to drifting off site.

dust containers, or equipment failure. Usually these accidents happen while someone is mixing chemicals or making an application (Figure 5-7). Accidents may result in large spills of dusts or of diluted or undiluted pesticide liquids, and a spill may be difficult to contain. The most important steps to take when an accident occurs are to

- protect people and animals in the area from contact with the spilled material
- contain the spill to prevent it from spreading (Figure 5-8)
- get help to clean up the spill (contact the county agricultural commissioner for resources), remove contaminated soil, and clean contaminated surfaces

FIGURE 5-5.

As the distance a droplet must travel to the target plant increases, the chances of drift increases. Never exceed the nozzle manufacturer's recommended nozzle height.

FIGURE 5-6.

Keep people out of the area during and after the application until the label instructions indicate that it is safe to enter.

How Pesticides Enter the Body

When pesticide exposure occurs, the pesticide can be absorbed into the tissues of the exposed person. Absorption can take place through four exposure routes: *dermal* (skin), *ocular* (eyes), *respiratory* (lungs), and *oral* (mouth) (Figure 5-9).

Dermal Exposure. Skin contact is the most frequent type of pesticide exposure. On contact with the skin, certain pesticides may cause a skin rash or mild skin irritation known as

FIGURE 5-7.

The most serious pesticide exposures are generally caused by accidents during mixing or application.

FIGURE 5-8.

It is important to contain a pesticide spill so it will not continue to spread during clean-up.

dermatitis. Other types of pesticides cause more severe skin injury, such as burns. Even internal poisoning may result if a pesticide is absorbed through the skin. In this case, the blood carries the absorbed pesticide to other organs inside the body.

The ability of a pesticide to penetrate skin depends on characteristics of the pesticide and its formulation. Oil-soluble pesticides are more likely to pass through skin than those that are soluble in water. In addition, skin on different parts of the body absorbs pesticides at different rates. For instance, the skin around the head absorbs pesticides many times faster than the skin of the hands or forearms.

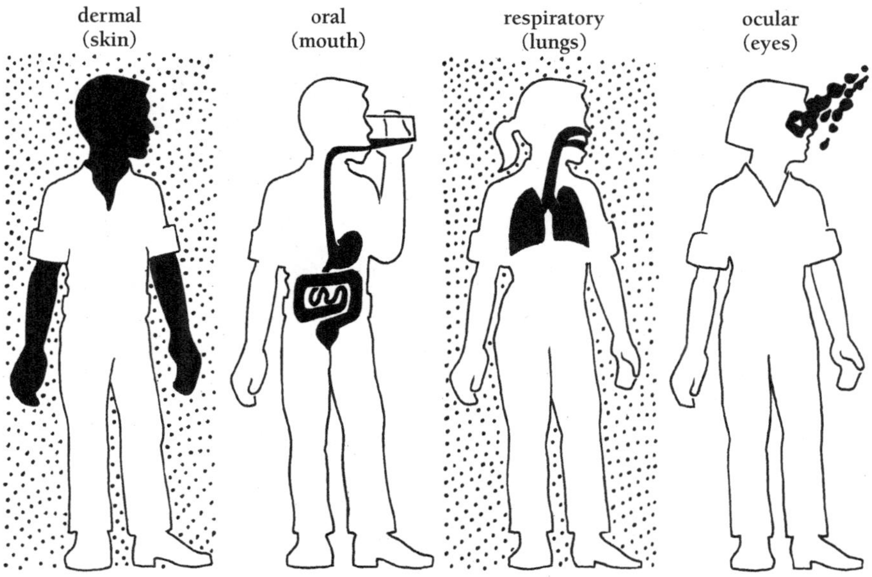

FIGURE 5-9.

The most common ways for pesticides to enter the human body are through the skin (dermal exposure), mouth (oral exposure), lungs (respiratory exposure), and eyes (ocular exposure).

Ocular Exposure. The active or inert ingredients of some pesticide formulations are caustic to the eyes. Besides being vulnerable to physical injury, the eyes are also very absorbent and so provide an easy route of entry into the body for certain pesticides.

Respiratory Exposure. The lungs quickly absorb certain pesticide dusts or vapors and provide a rapid route to the bloodstream, which can then transport the pesticide to other parts of the body. In addition to their potential for absorption, some pesticides also cause serious lung injury. It is difficult to avoid breathing dusts or vapors while mixing or applying these pesticides unless you use appropriate respiratory equipment. *Always* wear a label-recommended respirator during mixing and application. Make sure any respirator that you use fits properly and is in good condition.

Oral Exposure. The linings of your mouth, stomach, and intestines readily absorb pesticides. You can become very ill if even a small amount of pesticide comes into contact with your lips or the inside of your mouth, or if you swallow a small amount. Although it is rare for anyone to drink a pesticide accidentally, such a thing can happen if pesticides are improperly stored or negligently put into food containers. Children have been seriously injured in this way (Figure 5-10). *Never put pesticides into food or drink containers.* Keep all pesticides in their original packages. Do not mix or measure pesticides with utensils that someone could use later for food preparation or serving.

Oral exposure occurs more commonly when spray materials or pesticide dusts splash or blow into someone's mouth during mixing or application. People occasionally ingest pesticide residues when they eat or drink foods or liquids that have been contaminated with pesticide residues through drift or in some other way. Smoking while you handle pesticides increases the risk that you will inhale pesticides because tobacco readily absorbs pesticide materials. Before you eat, drink, or smoke, wash your hands thoroughly. Keep food and drinks away from areas where pesticides are being applied or mixed.

FIGURE 5-10.

Most pesticide poisoning victims are children. The improper storage of pesticides in food containers is the prime reason why children find and ingest pesticides.

Effects of Exposure

Immediate *(acute)* pesticide illnesses can result from a single exposure to a pesticide. In other cases, illness might occur after repeated exposure to small doses over time. It is common for different individuals to vary in their sensitivity to an amount of pesticide exposure. One person may show no reaction to a dose that might cause severe illness in another. A person's age and body size often influence his or her response to a particular dose. Thus, smaller doses might have more impact on infants and young children than they have on adults. Smaller doses may also be more harmful to the elderly than to younger adults, and adult females can be more sensitive to a particular dosage than adult males.

The type and severity of injury or poisoning that results from pesticide exposure depends on the pesticide's toxicity and mode of action. The amount absorbed and the speed of absorption also influences the severity of the injury. In many cases, the speed with which a person's body breaks down and excretes the pesticide affects the extent of injury. Prompt first aid and medical treatment may reduce the potential for injury.

Types of Injuries

Very small doses of some pesticides usually cause no injury or poisoning symptoms. Depending on the toxicity of the pesticide, a moderate or large dose may cause severe illness. Effects of exposure may be localized, such as irritation of the eyes, skin, or throat. Some effects may also be generalized if the pesticide absorbs through the skin. Once exposure occurs, certain pesticides may affect several different internal organs at the same time. The type and characteristics of the pesticide and the amount of exposure determine the level of damage.

A single massive dose absorbed during one pesticide exposure incident may cause injury, or in rare cases, death. Injury may also result from smaller doses absorbed during repeated exposures over time. The illness or damage may appear suddenly and last a short while or it may continue for a long time. Most injuries caused by pesticides are reversible—either the body will repair itself or medical treatment will cure the condition. Accidental exposure to a few types of pesticide, however, may cause irreversible, or permanent, damage. Sometimes this leads to an extended, chronic illness or disability. Severe exposure to certain pesticides can result in death.

There are many other factors that can produce symptoms that are similar to those of pesticide poisoning. Heat stress or excessive alcohol consumption, for example, can give you a headache or make you nauseous. Some plants cause skin irritation. If you are uncertain about the cause of your symptoms, seek medical attention if you have been in a situation where pesticide exposure *could* have occurred. If pesticide exposure is responsible for your symptoms, make every effort to locate the source of exposure and take action to prevent it from happening again.

Other Effects on People

Allergies. A few people may show signs of allergic reactions when exposed to some types of pesticide. The material actually causing the reaction could be the pesticide itself or one of the other components of the pesticide formulation. Allergic symptoms often include breathing difficulties, sneezing, watery and itchy eyes, skin rashes, anxiety, and general discomfort. Sensitive people should avoid exposure to pesticides that produce allergic reactions. To protect sensitive individuals, try different types of pesticide or different formulations, or switch to non-chemical control methods.

Anxiety. A number of people have serious concerns about the possible dangers that pesticides may pose to their health. Some may even develop symptoms of an illness if they suspect they were exposed. The illness is real, but there may have been no actual pesticide exposure or injury. Sometimes an odor can trigger this type of reaction. Usually, people associate pesticide odors with toxicity, believing that the stronger or more offensive the odor is, the more toxic the pesticide—but odor is not really an indicator of toxicity. For example, methyl bromide is extremely toxic but typically has no detectable odor at all. Other pesticides that have strong, objectionable odors may be far less hazardous.

OTHER HAZARDS

Injury to Pets

Dogs, cats, and other pets are common visitors to landscape areas (Figure 5-11). Some landscapes also include decorative ponds stocked with ornamental fish such as koi or goldfish. Like humans, animals can come into contact with pesticide residues after you have made an application. Unfortunately, it is difficult to keep animals out of treated areas or avoid pesticide drift into ponds where fish live.

Dogs and cats are usually at a higher risk of exposure to pesticide residues than people are, and may even contact sprays directly because of their habits, small size, and inability to understand the dangers. A pet that walks through a freshly treated area may bring pesticide residues into homes on its feet and coat. In an area where such animals may be present, make efforts to protect them and to prevent the spread of pesticide residues. Some steps to take include

- making applications at times when the animals are not present
- using spot treatments to avoid leaving large areas of residues
- notifying pet owners before the pesticide application
- selecting materials that are less hazardous to pets, ornamental fish, and wildlife
- leaving buffer strips around ponds, avoiding drift into pond water, and not applying pesticides in areas where irrigation runoff will flow into a pond (Figure 5-12)

FIGURE 5-11.

Dogs, cats, and other pets, like humans, can come into contact with pesticide residues after an application.

FIGURE 5-12.

Do not apply pesticides in areas where irrigation runoff flows into a pond.

- relying on non-chemical control methods whenever possible

Contamination of Surface Water

Several pesticides, including diazinon and chlorpyrifos, are contaminating streams, rivers, lakes, and bays in California, as well as the ocean, and are threatening aquatic animals (Figure 5-13). This contamination can also affect the quality of drinking water. These problems have prompted U.S. EPA and the State of California to

FIGURE 5-13.

Certain pesticides have contaminated California's inland streams, rivers, lakes, and bays as well as ocean waters and pose a threat to aquatic animals.

implement restrictions on many uses of the offending pesticides, including uses in landscape and turf. Other pesticides used in landscape and turf pest control may also contaminate surface water and in time their use may be restricted in a similar way.

In urban areas, pesticides can reach creeks and rivers through storm drains and sewage drains. Rain and runoff from landscape and turf irrigation run down the streets, through gutters, and into storm drains (Figure 5-14). The runoff then flows through pipes and directly into creeks, lakes, and rivers.

Any disposal of pesticides or of water that has been used to clean, Personal Protection Equipment (PPE) or application equipment into sewers may also contaminate surface water. Sewers transport wastewater from toilets, sinks, showers, and other sewage drains to treatment plants. Although wastewater treatment plants pass incoming water through a thorough treatment and disinfecting process before the water is released into open waterways, that treatment process does not detoxify pesticide chemicals.

For more information on how to keep from contaminating surface water with pesticides, see *Pesticides and Water Quality* on the UC IPM Web site, http://www.ipm.ucdavis.edu/WATER/U/index.html.

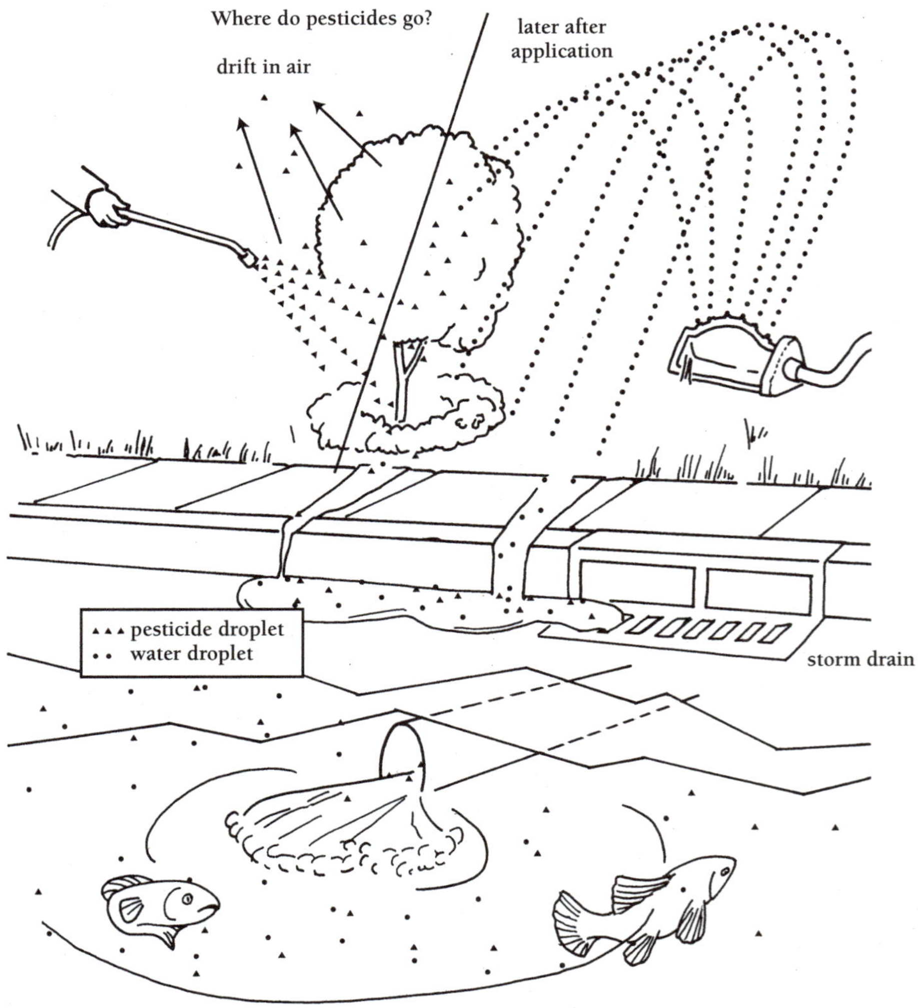

FIGURE 5-14.

Runoff from landscape or turf irrigation flows into storm drains, into pipes, and then into creeks, lakes, and rivers. If the runoff contains pesticides, it will contaminate these important waterways.

Groundwater Contamination

The potential for pesticide contamination of groundwater is a serious concern. Roughly 97% of the world's water is oceanic salt water and is unsuitable for most agricultural, industrial, and domestic uses. Less than 1% of all the planet's water is fresh water in an available form. People use this water for drinking, irrigation, household, and manufacturing purposes. Two-thirds of this fresh water is groundwater, trapped beneath the soil. The rest is surface water in lakes, ponds, streams, and rivers. Groundwater supplies nearly half of California's water needs. Cities obtain a large amount of their water from groundwater sources. Most of the water that people use in rural areas is groundwater.

A hazardous level of toxic chemicals in groundwater can affect large numbers of people. This concern has prompted lawmakers to develop stringent laws to protect groundwater, including laws regulating the use and disposal of pesticides.

How Groundwater Accumulates

After water passes through the soil surface, it slowly filters through soil particles until it reaches the groundwater basin—a process known as *percolation*. Common sources of percolated groundwater are rainfall, surface irrigation, and the melting of ice, snow, and glaciers. Some lake, pond, river, and stream water also percolates into groundwater basins, although most water from these sources flows out to the ocean. The amount and rate of percolation through a soil depend on the size of the soil particles. Fine particles such as those that make up clay and loam soils retain more water than do the larger particles that make up coarse, sandy soil. For this reason, percolation is slower in finer-textured soils.

Groundwater occupies large underground basins known as *aquifers*. Most aquifers are areas of water-saturated sand and gravel situated above or between one or more impermeable layers that confine the water. An aquifer may be hundreds of miles long and just as wide. It can range down to a depth as great as several thousand feet (Figure 5-15). Nearly half of California's landmass overlies groundwater basins. It is common to find several layers of aquifers beneath a single geographical area, with impermeable barriers separating each aquifer from others above or below it. When the barriers are broken through drilling or natural events, some of those separated aquifers can interconnect.

Like surface water, water trapped in an aquifer flows toward the sea, but it does so much more slowly than surface water. Water in a stream or river has a flow rate measured in feet per second, whereas underground water may only flow a few feet per year. Because the water moves so slowly, contamination may be difficult or impossible to contain. The water flows so slowly that it could take hundreds of years to remove contaminants from the system, even if appropriate technology existed.

How Pesticides Enter Groundwater

Pesticides enter groundwater in two ways: by leaching down through the soil or by entering directly through wells or other channels that connect with the aquifers.

Leaching Out of the Soil. Leaching occurs when rainwater or irrigation water percolates through the soil, dissolving water-soluble chemicals that may include some pesticides. In this way, dissolved pesticides can travel into the groundwater source.

Direct Entry through Wells or Other Structures. When there is backflow of pesticide-contaminated water into a well, it will contaminate the groundwater. This is why regulations require incorporation of an air gap or backflow device into the plumbing when you fill a spray tank from a well or any other

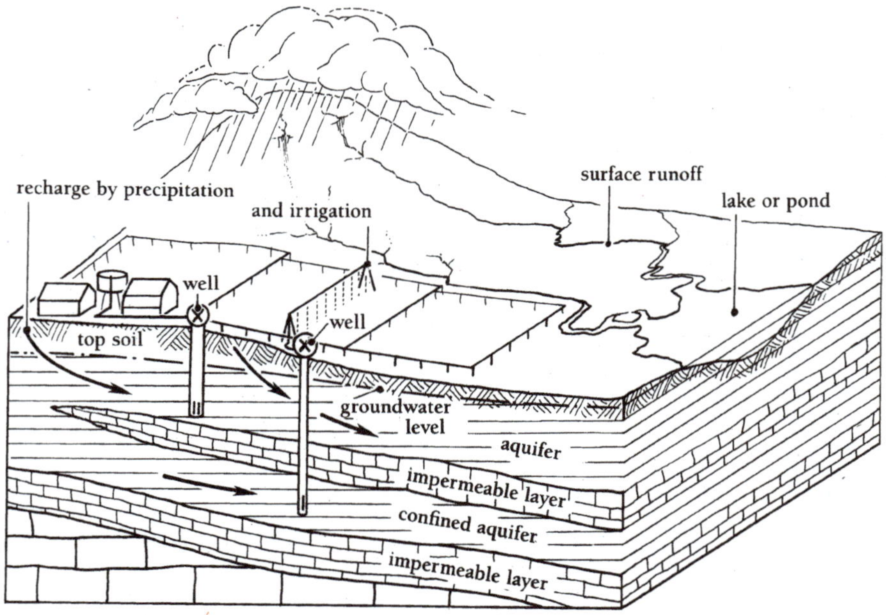

FIGURE 5-15.

Water is present in underground reservoirs, called aquifers. Sand and gravel formations contain this water while impermeable layers of clay, silt, or solid rock prevent water from leaving the aquifer.

water source. In addition, spills of pesticides or pesticide mixtures near a well may cause some of the material to flow down the well casing, contaminating the groundwater.

Factors Influencing Groundwater Contamination

Many factors influence a pesticide's potential for groundwater contamination. An important one is the chemical characteristics of the pesticide itself. For instance, pesticides used in the past included many chlorinated hydrocarbon compounds. Many of these compounds are highly persistent in the environment, but also immobile because they are not soluble in water. In addition, the chemical structure of some of these compounds causes them to bind tightly to soil particles.

Recent trends in pesticide use have been toward less-persistent chemicals, but many of these are highly mobile. These include some of the phenoxy, urea, and carbamate compounds that do not bind tightly with soil particles. Some are soluble in water, and so leach more easily.

Soil type affects a chemical's potential to contaminate groundwater. Geological formations such as rock layers beneath the application site also affect percolation. Liquids percolate more quickly through sand or gravel soils than through denser silts, loams, or clays. High amounts of organic material in the soil tie up a pesticide and may speed its breakdown. The following factors alter the stability of pesticides and affect their ability to reach groundwater:

- soil temperature
- pH
- moisture content
- dissolved salts
- quantity and type of soil organisms

There is less risk if no aquifer lies beneath the treatment area or if the aquifer is very deep. Shallow aquifers beneath treated areas are more susceptible to contamination because pesticides have to pass through less soil before reaching the aquifer and so have less opportunity to break down. In some locations, layers of impermeable subsoil prevent pesticides from leaching into the aquifer. The slope of the

TABLE 5-1.

Factors Influencing Pesticide Leaching.

CULTURAL PRACTICES
☐ amount and type of pesticide used
☐ method of pesticide application
☐ irrigation practices at treatment site, including frequency of irrigation and timing of irrigation in relation to pesticide application
GEOLOGIC CONDITIONS OF TREATMENT AREA
☐ slope
☐ underlying formations
☐ proximity of surface water channels such as ponds, lakes, and rivers
INTERACTION OF PESTICIDES IN SOILS
☐ properties of pesticides, including:
a. water solubility
b. volatility
c. soil adsorption
d. decomposition
☐ soil influence on pesticides, including:
a. soil texture
b. soil organic matter content
c. soil water content

treated land often determines whether contaminated water will run off or leach. Table 5-1 summarizes factors that influence groundwater contamination by pesticides.

Conditions that favor pesticide decomposition also reduce the risk of groundwater contamination. Research suggests that groundwater environments may slow the breakdown of some pesticides. For example, groundwater is not exposed to air or to ultraviolet light, both of which contribute to pesticide breakdown. Also, groundwater is cooler than the soil surface and lacks much of the biological activity that is found in soil and contributes to pesticide breakdown.

Injury to Plants

Phytotoxicity (chemical injury to plants) may be a problem when a given pesticide is used on certain landscape and interiorscape plants. The pesticide's active ingredient, however, is not always the cause of phytotoxicity. It may instead be caused by solvents or other inert ingredients in the formulation or by impurities (such as salts) in the water that is mixed with the pesticide. Excessive application rates, inadequate mixing, or improper pesticide dilution may also cause plant damage. Environmental conditions such as temperature and humidity levels at the time of application can often increase phytotoxicity. Also, plants that are stressed for water or nutrients may be more susceptible to injury.

Symptoms of Phytotoxicity. Phytotoxicity symptoms include the curling, cupping, twisting, yellowing, scorching, and dropping of leaves on plants as well as dead patches on turfgrass. Other types of non-pesticidal injury can cause some of the same symptoms, however. Before you come to a conclusion about the cause of symptoms like these, seek help from your local UC Cooperative Extension office. For more information on abiotic disorders, see *Abiotic Disorders of Landscape Plants: A Diagnostic Guide,* UC ANR Publication 3420. Also, refer to the online *UC Guide to Healthy Lawns: Abiotic Disorders* (http://www.ipm.ucdavis.edu/TOOLS/TURF/PESTS/abiotic.html) for phytotoxicity symptoms on turf that result from exposure to herbicides and other chemicals.

Damage to Objects in the Treated Area

Pesticides may sometimes cause spotting on treated plants or may damage the surface of vehicles, furniture, or other materials in the treated area. The spots may be superficial and you may be able to simply wash them off. Some pesticide residues, however, may cause permanent damage. A pesticide or the inert ingredients in a formulation may etch paint or other surface coatings or cause staining. The application rate and the spray mixture concentration can influence the occurrence and severity of spotting, pitting, staining, and other damage. To avoid this kind of damage, follow label instructions for mixing and application.

Impact on Nontarget Organisms

Some types of pesticide may be harmful to nontarget organisms at the application site and in the surrounding environment.

Natural Enemies and Other Beneficials

Natural enemies are insects and other organisms that feed on or parasitize pest insects and mites and help keep their populations low. Many of these insects and mites contribute significantly to the long-term natural control of plant pests. In addition, other non-harmful insects *(beneficials)* such as honey bees, wild bees, certain wasps, and other insects are important pollenators of fruit trees and other plants. Some beneficial insects help decompose dead plant and animal material. Beneficial bacteria, fungi, and nematodes also play important roles in decomposition.

When you kill natural enemies, you disrupt the natural control of pests and, often, increase your dependency on the use of pesticides. Elimination of natural enemies may cause pest resurgence or secondary pest outbreaks (Figure 5-16).

To protect natural enemies,

- eliminate the unnecessary uses of pesticides

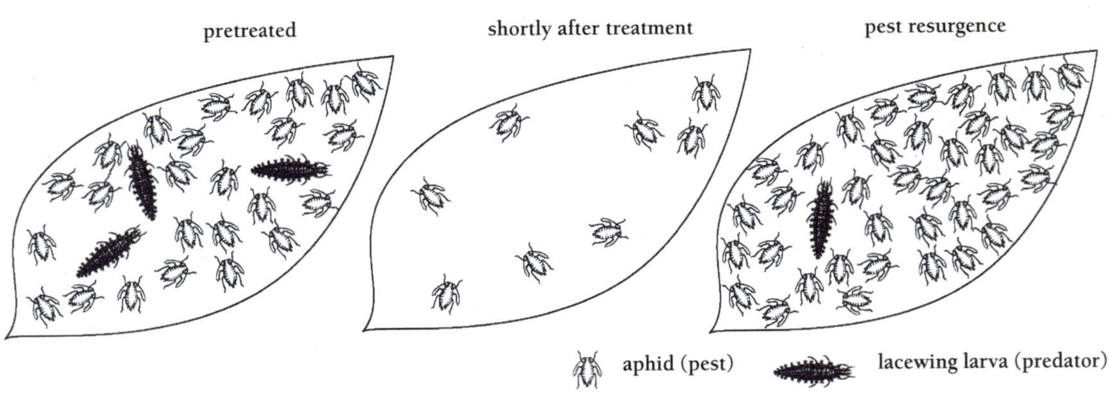

FIGURE 5-16.

The elimination of natural enemies may cause pest resurgence, as shown in this illustration. Although the pesticide application lowered the pest population at first, the pest population climbed to an even greater number soon afterward because the application also destroyed the natural enemy population.

- when pesticides are needed, choose pesticides that are less toxic to natural enemies
- apply pesticides at times when natural enemies are least likely to be harmed, such as during plant dormant periods
- use spot treatments wherever possible to minimize widespread effects
- use lower dosages when possible

Wildlife

Some pesticides can injure or poison and kill wildlife or cause them indirect harm by altering their food sources or habitats. Pesticides can cause serious harm to birds, fish, and other vertebrates. Certain animals are frequent unintended victims of baits used to control target pests. Although a pesticide may not directly cause the death of a nontarget animal, the chemical's effect might weaken the animal and so lead to illness or death because the animal is unable to get food and water or to protect itself from predators. Some pesticides may influence the ability of wildlife to reproduce.

CHAPTER 5: REVIEW QUESTIONS

1. Which of the following is the most frequent route of pesticide exposure among pesticide workers?
- ☐ a. oral (through the mouth)
- ☐ b. dermal (through the skin)
- ☐ c. inhalation
- ☐ d. eye

2. Which formulation offers the greatest potential for dermal absorption?
- ☐ a. water-soluble liquids
- ☐ b. powder formulations
- ☐ c. oil-soluble formulations
- ☐ d. granulars

3. Which of the following is not a factor in the soil-leaching ability of a pesticide?
- ☐ a. time of application
- ☐ b. the chemical nature of the pesticide
- ☐ c. soil type
- ☐ d. the pesticide's persistence

4. Which of the following practices will reduce environmental contamination by pesticides?
- ☐ a. Make frequent applications of various pesticides
- ☐ b. Avoid non-chemical control methods whenever possible
- ☐ c. Reduce the frequency of applications whenever possible
- ☐ d. Use maximum allowable rates of pesticides at all times

5. Not using a backflow device when filling spray tanks
- ☐ a. improves water quality in the spray tank
- ☐ b. is allowed in arid parts of the state
- ☐ c. may result in serious contamination of the water source
- ☐ d. speeds up the mixing process

6. Insect predators and parasites and honey bees are all considered
- ☐ a. secondary pests
- ☐ b. natural enemies
- ☐ c. beneficials
- ☐ d. pollinators

7. The seriousness of an injury caused by pesticide exposure usually is related to the
- ☐ a. time of day the exposure occurs
- ☐ b. toxicity and dose of the pesticide
- ☐ c. type of application equipment used
- ☐ d. frequency of application of that pesticide

8. Which of the following is the most important factor in pesticide drift?
- ☐ a. wind or other air movement
- ☐ b. droplet size
- ☐ c. spray pressure
- ☐ d. distance the droplets must travel

9. The most important step when a pesticide spill occurs is to
- ☐ a. direct the spilled material away from the area
- ☐ b. thoroughly soak the area to dilute the material
- ☐ c. protect people and animals in the area from contact with the material
- ☐ d. salvage as much of the spilled material as possible

10. The main reason food containers should not be used for measuring or storing pesticides is
- ☐ a. pesticides soften plastic and corrode metal
- ☐ b. people may mistake the contents for something to eat or drink
- ☐ c. these containers cannot be properly sealed
- ☐ d. they are not accurate enough for measuring pesticides

11. Which would be a factor influencing groundwater contamination?
- ☐ a. the chemical nature of the pesticide
- ☐ b. the speed at which the pesticide is applied
- ☐ c. spray pressure
- ☐ d. low wind speed during application

12. In addition to the pesticide active ingredient, etching of surfaces exposed to a pesticide application may be caused by
- ☐ a. pressure of the spray
- ☐ b. inert ingredients
- ☐ c. temperature of the spray solution during application
- ☐ d. size of spray droplets

13. One way to reduce pesticide damage to natural enemies and beneficial insects is to avoid
- ☐ a. using spot treatments
- ☐ b. spraying plants when they are dormant
- ☐ c. lower dosages when possible
- ☐ d. pesticides that are more toxic to beneficials

14. Which of the following could contribute to phytotoxicity problems when spraying a pesticide onto plants?
- ☐ a. adequate plant water and nutrients
- ☐ b. temperature and humidity at the time of application
- ☐ c. low application rates
- ☐ d. over-dilution of the pesticide mixture

15. Certain pesticides are more likely to cause groundwater contamination if they
- ☐ a. decompose quickly
- ☐ b. break down slowly
- ☐ c. volatilize easily
- ☐ d. are insoluble in water

6 Protecting People and the Environment

THERE IS A WIDE range of risks and a relatively limited range of benefits to using pesticides. When used properly, pesticides can effectively control pests and protect valuable plants or turf from damage. In many situations, pesticides may be the most cost-effective way to manage certain pests. The risks, as discussed in the previous chapter, include hazards to people, to the treated site, and to the environment. For these reasons, any decision to use pesticides in landscape, turf, or interiorscape situations must be well thought out. Whenever possible, look for alternatives to using pesticides to manage pests. For help choosing alternatives, consult the *Pest Notes* on the UC Statewide Integrated Pest Management Program's Web site (http://www.ipm.ucdavis.edu/).

This chapter provides information on steps you can take to protect people and the environment from pesticide exposure. For more in-depth information on this subject, refer to *The Safe and Effective Use of Pesticides, Second Edition*.

PLANNING FOR PESTICIDE SAFETY

People who apply pesticides must cultivate certain habits and personal traits in order to make effective and appropriate pesticide use decisions, apply pesticides properly, and supervise others who perform pesticide applications. These habits and traits include the ability to

- be methodical, patient, and disciplined
- always use good judgment
- not cut corners or rush through a job
- stay focused on the job at hand

If you follow the pesticide label and obey the laws and regulations that deal with pesticides, you will be able to avoid most problems. To assure that pesticide applications run smoothly, you should

- learn about hazards in the area where you will be applying pesticides
- read and follow the pesticide label
- obey the laws and regulations that apply to all aspects of pesticide handling, transportation, storage, and application
- develop and use safe work habits and assure that employees also follow these
- wear the required personal protective equipment (PPE)
- provide employees with the required PPE and ensure that they use it
- avoid practices that may expose people or animals to pesticide spray or residues in your work situation
- check your equipment to make sure it is functioning properly
- avoid the use of alcohol, drugs, and medications that cause drowsiness while handling pesticides

Good planning for pesticide safety helps to prevent accidents. For instance, you should find out as much as you can about the pesticides you use by studying Material Safety Data Sheets (MSDS) (see below) and pesticide labels (label information is available from the Department of Pesticide Regulation Web site at http://www.cdpr.ca.gov/). That way you can learn what dangers you may encounter and what precautions you need to take. Finally, plan what you will do if an accident happens. Sidebar 12 is a checklist for planning pesticide applications.

Learn about Site-Specific Hazards

Before you begin a pesticide application, familiarize yourself with the physical characteristics of the terrain, greenhouse, building, or other application site. The pesticide label and MSDS include important information about the hazards the material poses to you and other people, animals, and the environment. Carefully inspect the treatment area for conditions or objects that may affect the safety of the application:

- Check for organisms (such as plants or animals) that might be injured and objects and structures that the spray mixture might damage.
- Identify environmentally sensitive areas near the spray application site, such as streams, irrigation ditches, ponds, lakes, fountains, homes, schools, or parks.
- Look for potential hazards that might endanger the operator during an application: in outdoor areas look for ditches, embankments, steep slopes, electrical wires, and electric fences; in buildings look for electrical outlets, motors, pilot lights and other sources of sparks or combustion, stairs, and improperly ventilated spaces.
- Be aware of the potential for damage that could be caused by the movement of the application equipment and operator through the area, and plan ways to avoid this.
- Time the application for the optimum plant growth stage and avoid stages when beneficials may be harmed, such as when honey bees are foraging on flowers.
- Consider applying at the time of day when the pesticide application will be least disruptive to people using the area.
- Plan ways to protect or remove pets, other animals, or objects in the area.

Material Safety Data Sheets

A Material Safety Data Sheet (MSDS) provides a great deal of detailed technical information about the hazards of a pesticide and its physical and chemical aspects. An MSDS is not, however, considered to be a pesticide label or any part of the labeling, and an MSDS cannot substitute for a label to meet any labeling requirements. An MSDS may duplicate some limited amount of information found in a pesticide label. Much of the MSDS information is not considered to be application use directions or requirements. Instead, MSDS information is intended to provide guidance, especially for emergency personnel responding to a pesticide spill, fire, or injury. Nevertheless, this information may be very helpful as a supplement to pesticide handler training and for your own safety. Following is a list of the sections that you will generally find in an MSDS, with examples of the type of information in each section:

- *Product Identification*— includes the product name and code, chemical family and name, synonyms, and chemical formula
- *Hazardous Ingredients*— the active and inert ingredients identified by name and CAS (Chemical Abstracts Service Registry) numbers, exposure limits, and concentration by percentage
- *Physical Properties*— includes a description of the physical form, color, odor, boiling point, and melting and freezing points of the product

SIDEBAR 12

Checklist for Planning a Pesticide Application

Employees:

- ☐ Properly trained for this type of application?

Pesticide:

- ☐ Read and thoroughly understood label?
- ☐ Checked to be sure use is consistent with target pest and application area?
- ☐ Read MSDS (Material Safety Data Sheet) for information on hazards?
- ☐ Obtained necessary permits or operator identification number?
- ☐ Know proper rate of pesticide to be applied?

Equipment:

- ☐ Proper personal protective equipment (footwear, gloves, respiratory equipment, protective clothing, eye protection, headwear)?
- ☐ Necessary measuring and mixing equipment?
- ☐ Suitable application equipment for this job (tank capacity, pressure range, volume of output, nozzle size, pump suitable for formulation type)?
- ☐ Application equipment properly calibrated?
- ☐ Decontamination facility supplied with emergency water and first aid supplies?
- ☐ Necessary supplies to contain spills or leaks (absorbent materials, cleaning supplies, holding containers)?

Transporting:

- ☐ Can transport pesticides safely to application site?
- ☐ Pesticides and containers secured from theft or unauthorized access?
- ☐ Vehicles properly marked and permits obtained, if necessary, for transporting hazardous materials and hazardous wastes?

Mixing and Loading:

- ☐ Safe mixing and loading site located?
- ☐ Clean water for mixing available?
- ☐ Water pH tested?
- ☐ Proper adjuvants obtained to correct pH, prevent foaming, and improve deposition?
- ☐ Checked compatibility of pesticide tank mixes or fertilizer-pesticide combinations?
- ☐ Liquid containers triple-rinsed and rinse water put into spray tank?

Treatment Site:

- ☐ Boundaries of treatment site inspected?
- ☐ Environmentally sensitive areas within and around treatment area identified?
- ☐ Notified people working or living in or near treatment area?
- ☐ Treatment site properly posted with required signs?
- ☐ Soil types determined and noted, if these are a factor in pesticide efficacy?
- ☐ Pets and other animals properly protected?
- ☐ Aspects of groundwater determined, if applicable?
- ☐ Hazards within treatment site identified, including electrical wires and outlets, ignition sources, obstacles, steep slopes, and other dangerous conditions?
- ☐ Plants in treatment area in proper condition for pesticide application (correct growth stage, not under moisture stress, other requirements as specified on pesticide label)?

Weather Conditions:

- ☐ Weather suitable for application (low wind, proper temperature, lack of fog or rainfall)?

Application:

- ☐ Application pattern established that is suitable for treatment area, hazards, and prevailing weather conditions?
- ☐ Application rate selected that will give most uniform coverage?
- ☐ Equipment frequently checked during application to ensure that everything worked properly and provided a uniform application?

Cleanup:

- ☐ Application equipment properly cleaned and decontaminated after application?
- ☐ Personal protective equipment safely stored and then cleaned or laundered according to approved methods?
- ☐ Disposable materials disposed of in approved way?

Disposal:

- ☐ Plastic and metal containers triple-rinsed?

- ☐ Plastic and metal containers properly stored until disposed of in suitable disposal area?
- ☐ Paper containers properly disposed of?

Storage:

- ☐ Unused pesticides returned to supplier or stored in a locked facility for later use?
- ☐ Storage facility suitable for pesticides?

Reports and Records:

- ☐ Necessary reports filed with requesting agency?
- ☐ Necessary records completed and filed?

Follow-up:

- ☐ Treatment areas inspected after application to ensure that the pesticide controlled the target pests without causing undue damage to nontarget organisms or surfaces of items in treatment area?

Damage:

- ☐ Damage, if it occurred, promptly reported?

- *Fire and Explosion Data*— flash point, extinguishing media, special fire fighting procedures, and unusual fire or explosion hazards
- *Human Health Data*— routes of entry, human effects and symptoms of overexposure including acute and chronic effects of overexposure, a rating of the carcinogenicity of the product, and medical conditions aggravated by exposure
- *Emergency and First Aid Procedures*— first aid for eyes, skin, inhalation, and ingestion, and information for physicians, including antidote
- *Employee Protection Recommendations*— requirements for eye and skin protection, respirator and ventilation requirements, medical surveillance, and additional protective measures
- *Reactivity Data*— stability and instability conditions, hazardous polymerization (a chemical reaction where two or more small molecules combine to form a larger molecule), incompatibilities, and decomposition products
- *Spill and Leak Procedures*— procedures for dealing with spills and leaks, and disposal methods
- *Special Precautions and Storage Data*— minimum and maximum storage temperatures, shelf life, special sensitivity, and handling/storage precautions
- *Shipping Information*— technical shipping name; bulk and package freight class; domestic surface, ocean, and air shipping information
- *Animal Toxicity Data*— acute, sub-chronic, chronic, and developmental toxicity, carcinogenicity, mutagenicity, reproduction, and neurotoxicity data
- *Federal Regulatory Information*— OSHA, TSCA, and RCRA status, CERCLA reportable quantity, and SARA Title III information (see Glossary for definitions of these acronyms)

Pesticide manufacturers prepare the MSDS and make it available to every person selling, storing, or handling the pesticide. You can obtain them from the pesticide manufacturer, pest control dealer, or other supplier. You can get an MSDS for every registered pesticide. One internet resource for many pesticide MSDSs is http://www.cdms.net/manuf/manuf.asp. A downloadable MSDS should not, however, be seen as a substitute for the MSDS supplied by a chemical manufacturer.

Pesticide Labels

You must follow the directions on the label for any pesticide use. The label is a legal document that must be at the application site. The label gives specific information for each type of registered application. Thoroughly read and understand the entire label before you purchase the pesticide and before each use. This is most important since a pesticide manufacturer may change a pesticide label at any time. Changes may affect the registered use sites, PPE and other handler and worker safety requirements, and use directions. It is possible to have on hand multiple containers of the same pesticide that have different labels. If this happens, you must follow the label associated with the specific container that you are using. When you switch to another container, follow its specific label.

When reading the label, look for signal words (*Danger* or *Danger Poison*, *Warning*, or *Caution*). Check the label for required work clothing and PPE. Always use this equipment and make sure that it is in good condition. Review the label for any special environmental precautions. Make certain that the label lists the intended landscape or interiorscape application site. Finally, consult the label for information on how to dispose of unwanted pesticide and then compare this information with any county, regional, or state disposal requirements before you dispose of unwanted material.

Prepare to Respond to Accidents

Be prepared to deal with pesticide exposure incidents or accidental spills by planning for them. This includes

- having an emergency response plan
- locating an appropriate emergency medical facility before you need emergency care
- finding out in advance where to get assistance with spill cleanup
- maintaining an emergency spill cleanup kit
- posting in your vehicle the telephone number and location of a medical facility close to where you are working
- writing down the telephone numbers of the local fire department, sheriff, and highway patrol (the emergency number 9-1-1 usually gives you immediate access to medical help, local fire services, and law enforcement agencies)

If you have a pesticide spill, be ready to protect the public from danger. Know the proper first aid to administer to victims of pesticide exposure. This information is on the pesticide label and the MSDS.

Make sure that you have adequate clean water for washing your eyes and skin. State regulations require that sufficient water, soap, and single-use towels for routine washing of hands and face and for emergency eye flushing and washing of the entire body be available for employees. This water must be within 100 feet of the mixing site. Usually, water from a hose bib meets this requirement.

Should you have an accident involving pesticides spilled onto public roads,

- immediately alert either the California Highway Patrol, county sheriff, city police, or local fire department
- keep people and vehicles away
- never leave the scene of a spill until responsible help arrives

For advice on cleaning up spills, contact CHEMTREC (Chemical Transportation Emergency Center) at 800-424-9300. *Call this number only in the case of an actual emergency.* There are times when the person or company transporting the material cannot quickly and safely clean up a spill. In these cases, municipal agencies either provide or contract for cleanup of roadways within their jurisdiction. These agencies include CalTrans (on state highways) and designated city or county agencies, including some fire departments. See Sidebar 13 for information on emergency help for dealing with a spill.

Pesticide Worker Training

California and federal pesticide laws

set minimum standards of training for all employees who handle pesticides as part of their work. Elements of this mandatory training fall into three categories: using pesticides safely, emergencies and health, and general information.

Using Pesticides Safely. This part of the training should be specific to the pesticides that employees handle or are expected to handle. The training must include

- the importance of wearing clean work clothing daily
- how to handle, open, lift, and pour containers
- how to operate mixing, loading, and application equipment
- procedures for adequately rinsing containers
- disposing of containers
- how to confine spray to the target area
- how to avoid contamination of people, animals, waterways, and sensitive areas
- how to transport containers
- how and where to store containers, and procedures to follow when containers cannot be locked up
- the importance of washing hands thoroughly before eating, smoking, drinking, or using the bathroom after any pesticide handling activities and the need to shower thoroughly when finished working for the day
- how to launder work clothing
- information on reading and understanding pesticide labels and MSDSs, including the signal words,

SIDEBAR 13

Emergency Numbers for Pesticide Accidents and Spills

WHEN PEOPLE HAVE BEEN EXPOSED TO PESTICIDES:

- ☐ Dial 9-1-1 for emergency medical assistance. Notify the 9-1-1 operator that the problem is a pesticide exposure. Provide an accurate location and information on the type of pesticide involved.
- ☐ After obtaining medical treatment for exposed individuals, determine whether a spill has taken place. Follow instructions below for a spill.
- ☐ Contact the county agricultural commissioner's office to report the incident. Find the telephone number in the white pages of the local telephone book, listed in the government pages under the county where the accident occurred.

FOR A PESTICIDE SPILL ON STATE OR FEDERAL HIGHWAYS:

- ☐ Notify the local office of the California Highway Patrol and the local fire department (dial 9-1-1). Inform the 9-1-1 operator that a pesticide spill has occurred. Provide an accurate location and the type of pesticide.
- ☐ Contact CHEMTREC at 800-424-9300 for assistance in cleaning up a pesticide spill.
- ☐ Contact the State of California Office of Emergency Services. Usually a written report will have to be filed. Check the government pages of the local telephone directory under "State Government Offices" or contact the main office in Sacramento: Governor's Office of Emergency Services, 2800 Meadowview Road, Sacramento, CA 95832, (916) 262-1843.
- ☐ Contact the local agricultural commissioner's office. Find the telephone number in the government pages of the local telephone book, listed under the county where the accident occurred.

FOR A PESTICIDE SPILL ON LOCAL, CITY, OR RURAL ROADS OR ON PRIVATE LAND:

- ☐ Contact the local police or sheriff and the local fire department (dial 9-1-1). Inform the 9-1-1 operator that a pesticide spill has occurred. Provide an accurate location and the type of pesticide.
- ☐ Contact CHEMTREC at 800-424-9300 for assistance in cleaning up a pesticide spill.
- ☐ Report the spill to the State Office of Emergency Services.
- ☐ Contact the local agricultural commissioner's office. Find the telephone number in the government pages of the local telephone book, listed under the county where the accident occurred.

precautionary statements, first aid instructions, and application rate, mixing, and application instructions
- recognizing pesticide poisoning symptoms
- the types of personal protective equipment and respiratory protective equipment you must use and how to inspect that equipment for damage or wear
- the proper fitting and wearing of personal protective equipment
- the proper care and maintenance of required personal protective equipment and respiratory protective equipment

Emergencies and Health. This part of the training provides information to employees on steps they can take to reduce the chance of injury should they become exposed to pesticides. It also provides them with information on obtaining medical care. The training must include

- first aid and decontamination procedures
- the procedures for handling non-routine tasks or emergency situations such as spills, leaks, or fires
- the name, address, telephone number, and location of the clinic, physician, or hospital emergency room for treatment
- company policy for reporting injuries or illnesses and obtaining medical treatment
- information about the symptoms and long-term health effects resulting from overexposure to pesticides
- conditions or pesticides that require medical supervision, and general provisions for medical supervision
- how to recognize and avoid heat stress and what to do should it occur

General Information. The final part of the training describes the legal responsibilities of employers and their employees. Training for this section includes

- the applicable laws and regulations and the importance of compliance with these
- the rights of the employee to receive information regarding pesticides to which they may be exposed
- protection against firing or other retaliatory measures that could be taken against an employee in response to his or her exercise of employee safety rights
- location of and instructions for accessing documents that pertain to the company's Hazard Communication Program, the pesticide labels, Pesticide Safety Information Series sheets, pesticide use records, medical records, and other documents

California DPR publishes the *Pesticide Safety Information Series* leaflets as a training aid for employers. California regulations require that these documents be part of pesticide handler and fieldworker training. Download these leaflets from the DPR Web site at www.cdpr.ca.gov/docs/whs/psisenglish.htm (or for Spanish language versions, change "psienglish.htm" to psispanish.htm"). You may also obtain copies of these leaflets from your local county agricultural commissioner's office. DPR produces two sets of leaflets for different employers. The "A" series leaflets apply to agricultural operations. The "N" series leaflets apply to landscape maintenance firms and other non-agricultural operations such as right-of-way maintenance businesses and structural pest control businesses.

Avoiding Alcohol, Drugs, and Certain Medications

Alcohol, drugs, and certain medications cause drowsiness, impair judgment, and often influence a person's ability to apply pesticides safely. These substances may also alter the toxicity of a pesticide in case of exposure. If you are taking any medication, consult a physician before you handle, mix, or apply pesticides. Do not use alcohol or drugs before, during, or immediately after making a pesticide application.

COVERALLS, WORK CLOTHING, AND PERSONAL PROTECTIVE EQUIPMENT (PPE)

Employers are responsible for providing the PPE and necessary training on using this equipment to employees who handle pesticides as part of their work duties. They are also responsible for cleaning and maintaining the employee's PPE. When the pesticide label or regulation requires the use of respiratory equipment, the employer is responsible for providing training to employees on the proper way to fit and use it.

Personal protective equipment helps protect your body and clothing from pesticide exposure. Some PPE also protects your eyes or prevents you from inhaling pesticide vapors. PPE is only effective, though, if it fits correctly and you use it properly. Always keep it cleaned and maintained according to manufacturers' instructions. Wear clean chemical-resistant or waterproof gloves when removing PPE after a pesticide application in order to prevent the transfer of residues onto your body.

The greatest risk of pesticide poisoning comes when pesticides contact your skin. Oil-soluble pesticides pass through skin more quickly than water-soluble pesticides. In addition, some parts of your body absorb pesticide more quickly than others.

The way you apply pesticides also influences the exposure that you receive. Applying chemicals in an enclosed area is usually more risky than applying pesticides outdoors.

Follow the pesticide label and DPR regulations to determine what protective clothing and equipment you must use. California PPE requirements for landscape maintenance personnel can be found in DPR's *Pesticide Safety Information Series* leaflet N-1. In most instances, the use of safer equipment than the label requires will increase your protection. The next section provides general information on PPE. *The Safe and Effective Use of Pesticides, Second Edition* has detailed information on types of PPE and how to clean and maintain this equipment.

Bodywear

You have many choices of PPE or clothing. Your personal preference and the type of work you perform may influence the styles or materials of the bodywear you select. Whatever bodywear you choose, however, must conform to pesticide label or regulation requirements. Common styles include coveralls and aprons. Manufacturers produce protective clothing made from several types of material. Some materials offer more chemical resistance than others or are more suitable for specific types of pesticides. Usually you have a choice of weight and strength of the fabric and of fabric resistance to ripping and puncturing. Different materials also provide options in comfort and durability. No single material provides everything, so select bodywear for your most important need: *protection from pesticide exposure.*

If the label or regulations require that you use waterproof or chemical-resistant bodywear, regulations impose certain temperature restrictions. Employees cannot wear waterproof PPE to apply these pesticides if the daytime temperature exceeds 80°F (26.7°C) or the nighttime temperature exceeds 85°F (29.4°C).

Minimal Protective Clothing. Never handle, mix, or apply pesticides without wearing at least some minimal work clothing to keep chemicals from contacting your skin. Refer to the label and regulations to determine the work clothing required for handling. *Work clothing* in this context means garments such as long-sleeved shirts, short-sleeved shirts, long pants, short pants, shoes, and socks. Work clothing is not considered PPE, although pesticide product labeling or regulations may require that you wear specific work clothing during some activities. Work clothing differs from and should not be confused with a coverall. While coveralls must be provided by the employer, an employer can require that work clothing be provided by the employee. Short-sleeved shirts and short pants are considered acceptable work clothing for pesticide handling only under conditions that are expressly permitted by pesticide product labeling.

Personal protective equipment means apparel and devices worn to minimize human body contact with pesticides or pesticide residues. PPE must be provided by an employer and is separate from, or in addition to, work clothing. PPE may include chemical-resistant suits, chemical-resistant gloves, chemical-resistant footwear, respiratory protection devices, chemical-resistant aprons, chemical-resistant headgear, protective eyewear, or a coverall (one- or two-piece garment).

The minimal protection required under low-hazard circumstances would generally be work clothing that includes a one- or two-piece garment of closely woven fabric (or equivalent) that covers the entire body except the head, hands, and feet (Figure 6-1). If you wear coveralls over your work clothing, you have additional protection from pesticides. Coveralls have the advantage of being easily removed if they become contaminated or when you finish working (Figure 6-2).

Disposable Protective Clothing. Manufacturers produce disposable protective clothing made of several types of material and suitable for pesticide application. Disposable fabrics made

FIGURE 6-1.

Minimal work clothing for low-hazard pesticides includes one- or two-piece garments of closely woven fabric that cover the entire body except the head, hands, and feet. Protective eyewear and chemical resistant gloves are also advisable gear.

FIGURE 6-2.

Coveralls are suitable for many types of pesticide applications. They have the advantage of being easy to remove if contaminated or when work is finished.

from nonwoven, bonded fiber materials are superior to those made from woven fabrics. This is because nonwovens do not promote wicking and are more resistant to liquid penetration. Uncoated disposable clothing is not suitable if the pesticide label states that the clothing must be "waterproof" or "chemical resistant," but manufacturers laminate or bond some nonwoven fabrics to other materials to make them waterproof. Disposables usually are lightweight but remarkably strong and resistant to tearing or puncture (Figure 6-3). Disposables have the major advantage that they do not require cleaning or decontamination after use. Just throw them away when soiled unless they are

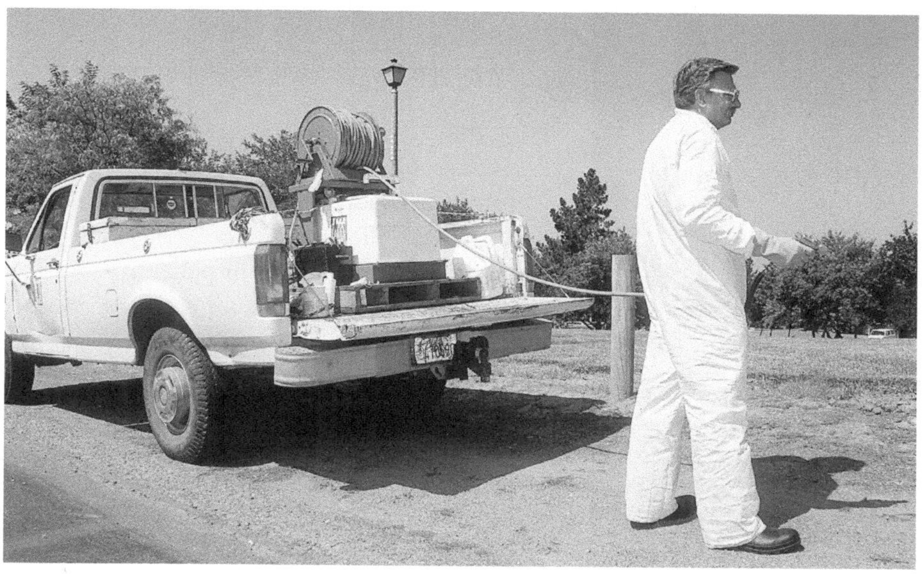

FIGURE 6-3.

Disposable protective clothing is inexpensive and lightweight and it resists puncturing and tearing. If it is not contaminated with concentrated pesticide, throw it away after use. Unless the fabric is specially coated, it is unsuitable for use when the label calls for a waterproof or chemical-resistant garment.

contaminated with undiluted pesticide. Dispose of any clothing contaminated with undiluted pesticide at a Class 1 disposal site (check with your county agricultural commissioner's office for information on Class 1 disposal sites).

Reusable Protective Clothing. Most reusable protective clothing is made of woven or nonwoven fabrics laminated to waterproof materials. The amount of protection offered against pesticide exposure depends on the type of waterproofing material. Neoprene, latex rubber, and polyvinyl chloride (PVC) all are very effective (Figure 6-4). Thin plastic material can tear or puncture easily and is not suitable for pesticide handling. Protective clothing is effective only so long as it is free from pesticide contamination on the inside and is in good condition. Replace or repair protective clothing when it becomes worn or develops tears or holes.

After use, remove as much pesticide residue as possible from reusable protective clothing. Before you take the garments off, wash them outdoors with a hose and scrub brush, and in an area where the runoff will not cause contamination. After you take them off, store them in a clean plastic bag until you can launder them. Follow the procedures listed in Sidebar 14 for cleaning protective clothing.

Aprons. An apron protects the front of your body during pesticide mixing and while you handle open pesticide containers (Figure 6-5). Do not use an apron as a substitute for other protective clothing during pesticide application unless the label or regulation permits such use. Read the pesticide label and follow DPR regulations to determine when to wear a protective apron (see DPR's *Pesticide Safety Information Series* leaflet N-1). Use an apron made of waterproof material. Make sure it is long enough to protect your clothing. Wide-bib styles provide splash protection for the upper chest. Disposable aprons made for one-time use generally are not suitable for pesticide handling. The thin plastic material tears and punctures easily. Reusable aprons are more durable, but must be cleaned and decontaminated regularly

FIGURE 6-4.

Chemical-resistant garments provide the maximum amount of protection from pesticide exposure.

FIGURE 6-5.

Refer to pesticide label instructions to see when it is safe to use waterproof or chemical-resistant aprons. Select a style with a wide bib to provide added protection.

SIDEBAR 14

Techniques for Washing Pesticide-Contaminated Clothing

1. Keep pesticide-contaminated work clothing separate from all other laundry.
2. Do not handle contaminated work clothing with your bare hands: wear rubber gloves or shake the clothing out from a plastic bag into the washing machine.
3. Wash only small amounts of work clothing at a time. Do not combine multiple pieces of work clothing contaminated with different pesticides—wash these in separate loads.
4. Before washing, presoak clothing:
   a. Soak in a tub or an automatic washer, or spray garments outdoors with a garden hose.
   b. Use a commercial solvent soak product or apply a pre-wash spray or liquid laundry detergent to soiled spots.
5. Wash garments in a washing machine, using the hottest water temperature, full water level, and normal (12-minute) wash cycle. Use the maximum recommended amount of *liquid* laundry detergent. Neither bleach nor ammonia seems to affect the removal of most pesticides. **Never use both!**
6. If garments have a pesticide odor, visible spots, or stains after washing, rewash them one or two more times as in step 5.
7. Clean the washing machine before you use it for other laundry. To do this, repeat step 5 with an empty load, but still using the full amount of hot water, normal wash cycle, and the maximum recommended amount of liquid laundry detergent.
8. Hang the cleaned garments outdoors on a clothesline so as to avoid contaminating your clothes dryer.

Do not attempt to wash heavily contaminated work clothing; dispose of it instead by transporting to an approved disposal site.

The following measures will reduce the chances that you will contaminate the family laundry and endanger your family members:

1. Whenever possible, wear disposable personal protective equipment that you can throw away after use.
2. Always wear all required personal protective equipment when you work with pesticides.
3. Begin each day with a clean set of personal protective equipment when you are working with pesticides. Wash all contaminated work clothing before you use it again.
4. Take off any contaminated work clothing at the work site and empty all pockets and cuffs. Place the clothing in a clean plastic bag until it can be laundered. Keep contaminated work clothing separated from all other laundry.

Remove any work clothing and personal protective equipment immediately if a pesticide concentrate has spilled onto it.

by washing in a detergent solution and then air-drying. Discard any apron that develops tears or holes.

Head Protection

Certain spray activities in landscapes and interiorscapes, such as spraying overhead, may expose your head to spray mist or droplets. To protect your head during these applications, wear a water-resistant, wide-brimmed hat. Headbands and sweatbands must also be waterproof. An alternative to a wide-brimmed hat is a hooded, waterproof jacket. Cloth baseball caps and similar hats are not suitable because they can absorb pesticide residues.

Gloves

A set of chemical-resistant gloves is an essential part of your safety equipment and California regulations require that employees wear them when engaged in virtually all pesticide handling activities. Never use leather or fabric gloves (unless specified on the pesticide label): they absorb water and pesticides. Choose unlined gloves made from nitrile, natural rubber, latex, butyl, or neoprene (Figure 6-6). The thickness of the glove material also influences the amount of protection—thicker materials are better. Choose materials that resist puncture and abrasion. Select unlined gloves since the

FIGURE 6-6.

Use only unlined chemical-resistant gloves made of natural rubber, latex, butyl, neoprene, or nitrile.

fabrics used for linings may absorb pesticides. When you are selecting gloves, consider comfort and mobility to allow dexterity. Check pesticide labels for special glove recommendations.

Footwear

Labels of some pesticides require the use of waterproof footwear or boot coverings (Figure 6-7). Select protective footwear made from natural rubber or synthetic materials such as PVC, nitrile, neoprene, or butyl. Choose the material based on its ability to protect you from the pesticides that you work with (Table 6-1). Common footwear made from leather, cloth, or synthetic materials absorbs pesticides and does not provide adequate protection from pesticide exposure.

Eye Protection

Always wear eye protection while handling pesticides. It is essential during mixing and loading and while adjusting, cleaning, or repairing contaminated equipment. In California, regulations require that employees use protective eyewear during virtually all pesticide handling activities, even if the requirement is not on the pesticide label. The regulation exemptions in which eye protection is usually *not* required are

- if pesticides are being injected or incorporated into the soil
- if pesticides are being applied through vehicle-mounted spray nozzles that are located below and behind the operator, with the nozzles directed downward
- if the operator is working in an enclosed cab
- if the pesticides being applied are rodenticides, predacides, or avicides that are not in a liquid or gaseous form

Eye protection is available in the form of safety glasses, goggles, face shields, and face shields that are combined with respirators. Unless the pesticide label specifies the type of eye protection, you may wear safety glasses if they include a brow piece and side shields as illustrated in Figure 6-8.

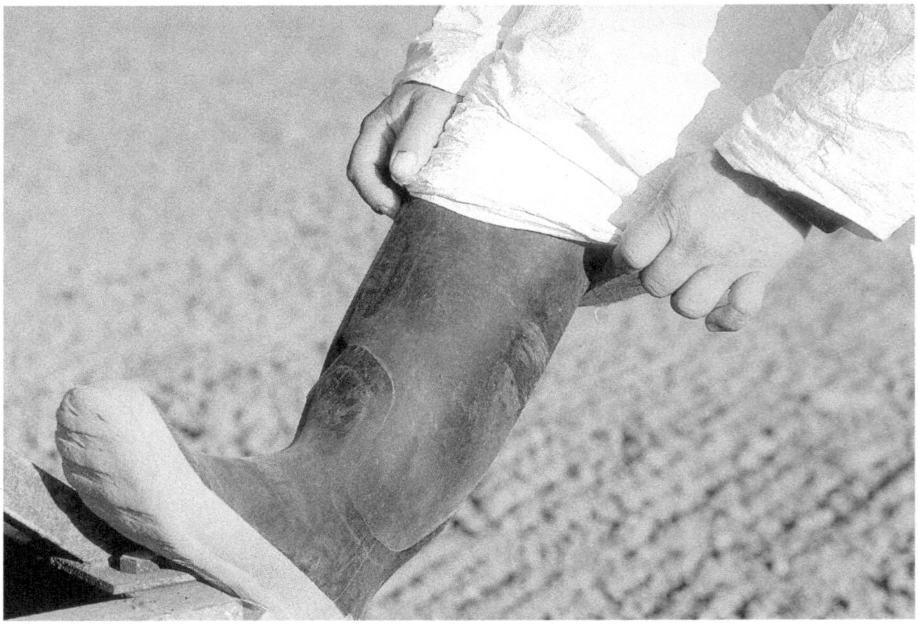

FIGURE 6-7.

Protective footwear is available in many styles, including an overshoe style to wear over regular shoes or boots. Whenever possible, clean your boots at the application site and before you remove them.

TABLE 6-1.

Guide to Selecting Chemical-Resistant Materials for Personal Protective Equipment.

PESTICIDE TYPE OR COMPONENT	PVC	NITRILE	NEOPRENE	BUTYL RUBBER	STANDARD RUBBER	LEATHER	FABRIC
Alcohols	●●	●●	●●	●●	●	♦	♦
Hydrocarbon materials, oils, and solvents (most pesticides)	●	●●	●	●	♦	♦	♦
Inorganic metals	●●	●●	●●	●●	●●	♦	♦
Ketones and aldehydes	❍		●	●●	●	♦	♦
Organic esters	❍		●	●●	●	♦	♦

●● Highly satisfactory
● Often acceptable
❍ Occasionally acceptable
♦ Do not use

Respiratory Equipment

Respirators protect the lungs and respiratory tract from airborne pesticides. You can choose from several types and styles suitable for mixing and applying pesticides. Select respiratory equipment based on the requirements listed on the pesticide label. For more information on selecting, using, and maintaining respirators, see *The Safe and Effective Use of Pesticides, Second Edition.*

PUBLIC AND ENVIRONMENTAL SAFETY

Protect the public from exposure to pesticides during an application. It is your responsibility to keep members of the public away from treated areas during and after you make the application and until the spray dries or other entry restrictions have passed. To protect the public,

- notify people in the area about the planned pesticide application (some cities and schools may require that anyone applying pesticides notify people who use the area where the application will be made)
- explain the potential hazards and possible exposure symptoms
- describe what you are doing to reduce these hazards
- explain how people can avoid exposure
- make pesticide applications at times when people are not present
- prevent drift or other off-site movements of pesticide materials outside the treatment area.
- prevent water used for irrigation or other purposes from running off of pesticide-treated areas

FIGURE 6-8.

Eye protection is required for employees who perform most pesticide handling activities. Unless the label specifies the type of eyewear, you may use safety glasses that have a brow piece and side shields when handling pesticides.

Posting Treated Areas

On public property, California law requires that all applications of pesticides having worker restricted-entry intervals of at least 24 hours shall be posted with warning signs. A barrier that excludes people from the treated area may be used in place of the posted signs. The agency with responsibility for the property where the pesticide application is to take place is responsible for posting these signs or installing the barrier. Public property includes school grounds, parks, and other public rights-of-way where public exposure is foreseeable. The information on the signs must be written in English and Spanish and must contain a warning that the area has been treated with a pesticide and that individuals are not to enter.

- Post signs immediately before pesticide applications and remove them within 24 hours after the expiration of the restricted-entry interval.

- Place signs at regular intervals around the treated area and at all regular points of public entry.

HANDLING PESTICIDES SAFELY

Concentrated, undiluted pesticides are a greater risk to people and the environment than are diluted mixtures. The safe handling and transportation of undiluted pesticides can prevent many environmental and human health hazards. Besides their toxic hazards, pesticides have a high dollar value and may be subject to theft. Part of your safe handling program should include security measures to prevent theft.

Transporting Pesticides

Many pesticides are subject to state and federal hazardous materials transportation requirements, including requirements that cover

- incident reporting
- packaging (including container maintenance and re-testing)
- labeling
- marking
- placarding
- emergency actions in the event of an accident
- loading
- vehicle safety equipment
- routing

There is always a risk of an accident when you use a vehicle to transport undiluted or diluted pesticides. Spilled materials may cause serious human exposure as well as environmental damage. In addition, there are risks that passing vehicles may scatter any pesticides that are spilled on public roads, further endangering people. Chemicals that spill onto roads may wash into ditches, streams, and rivers during rainstorms and create the potential for serious environmental damage, including surface water and groundwater contamination. Pesticide spills may also contaminate the vehicle, its occupants, or other cargo, so a good practice is to place pesticide containers inside a secondary containment in the cargo area of the vehicle. A secondary

FIGURE 6-9.

Transport pesticides only in the back of a vehicle. Secure all pesticide containers in the cargo area and protect them from moisture and damage. Never carry people, animals, food, animal feed, or clothing in this cargo area when you are transporting pesticides.

containment could be as simple as a plastic pan that will confine any spilled material.

Regulations prohibit transportation of pesticides in the passenger compartment or other areas occupied by people in any vehicle. The safest way to transport pesticide material is in the back of a truck (Figure 6-9). Secure all pesticide containers in the cargo area and protect them from rain and other potential causes of damage. Never stack pesticide containers higher than the sides of the transporting vehicle. When transporting any pesticides, never allow children, adults, or animals to ride in the same cargo area. Also, never transport food, animal feed, or clothing in the same compartment. Never leave pesticides or pesticide mixtures unattended in a vehicle unless they are inside a locked compartment. To assist emergency workers in case of an accident, carry a manifest or list of the pesticides that are being transported in the vehicle.

For information on transporting pesticides on public highways, contact the California Highway Patrol and the U.S. Department of Transportation. These agencies will provide you with current regulations and help you determine if you need special licenses or permits. They will tell you if these materials are subject to hazardous materials transportation requirements and whether your vehicle may require placards that signify the type of hazardous material being carried. Sidebar 15 lists contact information for these two agencies.

Storing Pesticides

Store every pesticide in its original, tightly closed container. Make sure to clean any pesticide residue from the outside of containers using water and disposable towels. Protect pesticides from temperature extremes and from becoming wet. Your pesticide storage area should be a designated, separate building, away from people, living areas, food, animal feed, seed, and animals. The area must have good ventilation and lighting. Make sure it is dry and secure, with lockable doors and windows. To warn others that the building contains pesticides, post signs on structures near all primary entrances so they are visible from any direction of likely approach (Figure 6-10). Do not store anything else except other compatible pesticides in the facility. Never keep PPE in any pesticide storage area.

Keep records of all pesticides being stored. Indicate on these records the date of purchase and the date you placed each chemical into the storage area. Keep this record, or a copy of it, at a separate location away from the storage building. To prevent injury during a fire or other emergency, you have to give this information in advance to emergency workers. Before an emergency arises, provide the local fire protection agency with an inventory of the pesticides being stored. Also provide them with relevant MSDSs. Update your records periodically and whenever you add additional types of pesticides to the storage area.

Check stored pesticides on a regular basis. Inspect the condition of containers and look for leaks or spills. Immediately clean up any spilled pesticide in the storage area.

Some pesticides do not store well for long periods. This is because the chemicals in some pesticide formulations separate out over time. Also, prolonged storage may reduce the effectiveness of some pesticides, while it may make other pesticides more toxic. Pesticides in unsealed containers are most susceptible to these effects, which may be caused by temperature extremes, moisture, and exposure to air. Besides chemical-based changes, solvents and petroleum-based chemicals cause some types of containers to deteriorate after a time, creating leak hazards.

SIDEBAR 15

Where to Get Information on Transporting Pesticides

CALIFORNIA HIGHWAY PATROL

http://www.chp.ca.gov

☐ **Commercial Vehicle Section**
444 North 3rd Street, Suite 310
Sacramento, CA 94298-0228
Phone: (916) 445-1865
http://www.chp.ca.gov/html/cvs.html
http://www.cp.ca.gov/html/cvs.html

☐ **Motor Carrier Safety Program**

Northern Division
2485 Sonoma Street
Redding, CA 96001-3026
Phone: (530) 225-2715

Valley Division
1136 Trade Center Drive
Rancho Cordova, CA 95741
Phone: (916) 464-2090

Golden Gate Division
1551 Benicia Road
Vallejo, CA 94591-7568
Phone: (707) 648-4180

Central Division
4771 West Jacquelyn Avenue
Fresno, CA 93722-6406
Phone: (559) 445-6992

Southern Division
437 North Vermont Avenue
Los Angeles, CA 90004-3590
Phone: (323) 644-9557

Border Division
9330 Farnham Street
San Diego, CA 92123-1284
Phone: (858) 637-7158

Coastal Division
4115 Broad Street, Suite B-10
San Luis Obispo, CA 933401-7963
Phone: (805) 549-3261

Inland Division
847 East Brier Drive
San Bernardino, CA 92408-2820
Phone: (909) 383-4811

US DEPARTMENT OF TRANSPORTATION

☐ **Federal Motor Carrier Safety Administration (FMCSA) http://www.fmcsa.dot.gov**

For Northern California:
980 9th Street
Sacramento, CA 95814-2724
Phone: (916) 930-2760

For Southern California:
3401 Centrelake Drive Suite 550A
Ontario, CA 91761
Phone: (909) 937-2949

☐ **Federal Transit Administration**
201 Mission Street, Suite 2210
San Francisco, CA 94105-1839
Phone: (415) 744-3133
http://www.fta.dot.gov

Keep containers off the ground to reduce their exposure to moisture and regularly inspect containers for leaks or damage. Manufacturers recommend that you not store most pesticides longer than two years, so make sure not to overbuy pesticides and make sure to use the pesticides that you do buy in an appropriate application before they exceed their shelf life. If you have no planned use for stored pesticides, avoid potential storage problems such

FIGURE 6-10.

Store pesticides in a designated, separate building, away from people, living areas, food, animal feed, seed, and animals. Make sure the storage area is well ventilated, well lighted, dry, and secure. Securely lock doors and windows. Post the primary entrances with signs that warn that the building contains pesticides.

as leaking containers by having them transported to an approved disposal site.

Certain pesticides, such as 2,4-D and other hormonal herbicides, should not be stored with other pesticides. Vapors from these herbicides may combine with and contaminate the other pesticides and could result in potential damage to treated plants when other pesticides are applied. To prevent such problems, always check the pesticide label for special storage precautions.

Liquids expand when heated or frozen. Liquid pesticides in sealed containers may expand enough to rupture their containers under extremely hot or cold conditions. For this reason, you need to keep the storage area well ventilated to prevent overheating. When storing pesticides in an area subject to winter freezing, make sure to insulate the storage building.

Mixing Pesticides

Before you begin to mix any pesticide, determine what personal PPE you will need and what type of pesticide handling equipment you will require. Also, check your mixing, loading, and spray equipment for cracked hoses or leaks. Make sure the filters, screens, and nozzles are clean. You must have a supply of soap, single-use towels, and sufficient water (of a proper temperature and quality to prevent injury or illness) for routine washing and, in case of an accident, for emergency eye flushing and washing of the entire body.

Techniques for mixing pesticides are the same for large and small volumes—thoroughly mix the proper amount of pesticide with a measured amount of water or other diluent. Before you begin, read the mixing directions on the labels of all pesticides that you will be using. Then choose the proper order in which to add chemicals to the spray tank. Add formulations to the spray tank in the following order:

1. wettable powders
2. flowables
3. water-soluble concentrates
4. emulsifiable concentrates

If the mixture needs adjuvants, add

these before you add the pesticides unless label instructions specify a different sequence.

The water used for filling a spray tank should be clean and free of sand, dirt, algae, and other foreign matter. Sand and dirt cause excessive wear on pumps and nozzles and clog filters, screens, and nozzles. Algae may clog filters and nozzles and can react with some pesticides to reduce their effectiveness. High levels of dissolved salts deactivate certain pesticides and may even damage treated foliage.

Measuring. Measure pesticides carefully, accurately, and safely. Inaccurate measuring can cause serious problems such as the application of too much pesticide. This could result in damage to treated plants or harmful exposure to people and animals that may enter an area after the application.

Liquids and some granular pesticides are measured by volume, while dusts, powders, and most dry formulations are measured by weight. Most pesticide labels use the English system of measurement (fluid ounces, pints, quarts, and gallons for liquids; pounds and ounces for dry materials). You will need an assortment of glass measuring vessels to accurately measure liquids. Some pesticides bind to plastic or react with metals, especially aluminum and iron, so avoid using plastic or metal measuring vessels. Use a medicine dropper to measure small quantities of liquids. Use an accurate scale and a set of measuring cups and spoons for measuring and weighing dry pesticides (Figure 6-11). To avoid mistaking any pieces of measuring equipment for kitchen utensils, identify them in an obvious manner. For instance, paint the spoon and cup handles with brightly colored waterproof paint or attach waterproof labels to them.

When not in use, keep all measuring and weighing equipment locked in the pesticide storage area. This prevents it from being used for other purposes. Clean and wash utensils before storing them to prevent contamination of future mixtures.

Select a mixing location that is well lit so you can read all labeling or other written directions. There should be adequate lighting for you to safely perform mixing tasks and to thoroughly clean the area should a spill occur. When not using premeasured packets, measure and weigh chemicals in a clear, open area. If outdoors, stand upwind of the materials you are measuring to reduce the chance of exposure. Wear an approved dust-mist respirator or cartridge respirator while weighing and mixing dry pesticides to prevent inhalation of dust. Protect your hands and clothing with appropriate outerwear. Because liquids spill and splash easily, wear suitable gloves and a protective apron or a waterproof suit. Refer to the pesticide label for specific PPE required for mixing and loading the pesticide. All employees who are mixing, loading, or applying pesticides must always wear a face shield, goggles, or other protective eyewear, even if the requirement is not on the pesticide label (Figure 6-12). Reduce the chance that chemicals will spill or splash into your face and eyes by measuring and pouring pesticides below eye level.

Begin mixing by filling the spray tank at least half full with clean water. To allow room for the pesticide, adjuvants, and residues from triple-rinsing the empty containers, avoid filling the tank more than three-quarters full. Check and adjust the pH of the water in the spray tank at this time. See Sidebar 7 in Chapter 3 for instructions on checking the water's pH. Start agitators if the equipment has them.

Open all pesticide containers carefully to prevent spilling and to make resealing easier. Cut paper containers open with a sharp knife or scissors rather than by tearing. Metal and plastic containers all have protective seals that you must break before use.

FIGURE 6-11.

Measuring and weighing pesticides requires a variety of calibrated utensils and an accurate scale.

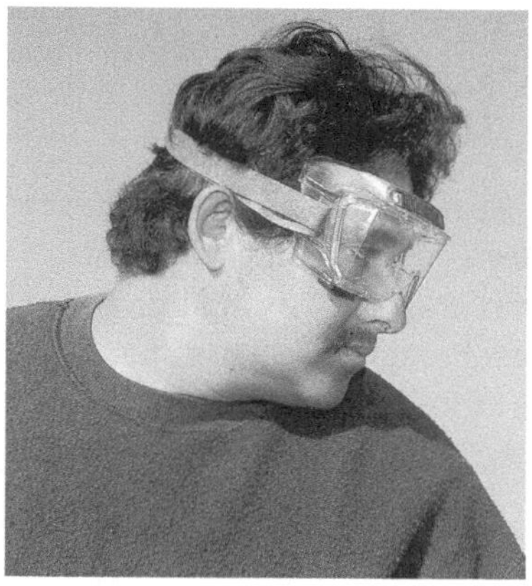

FIGURE 6-12.

Always pour and measure pesticides below eye level. If measuring outdoors, stand upwind of the chemicals. Wear the label-mandated PPE for mixers when measuring pesticides. California law requires that protective eyewear and gloves be worn by employees even if the label does not require them.

Designate a knife or pair of scissors for this purpose and keep it with your measuring utensils. Clean the knife or scissors after each use. If the seal becomes detached from a container, make sure to wash pesticide residue off it before throwing it away. Wash the seal over the spray tank's fill hole so the residue goes into the tank. Most metal and plastic pesticide containers have screw caps that allow you to reseal them easily.

After measuring or weighing the correct amount of pesticide, carefully pour it into the partially filled spray tank (Figure 6-13). Rinse the measuring container and pour the rinse solution into the spray tank. Use caution to prevent splashing while you rinse.

How to Triple-Rinse Liquid Pesticide Containers

Drain any liquid container into the spray tank for 30 seconds after you empty it. Rinse and drain the container at least three times (triple-rinse). After each draining, fill the container about one-quarter full of water and screw the cap on again. Shake the container for several seconds to mix any remaining residue with water. Pour each rinse solution into the spray tank. Containers that have been triple-rinsed do not have to be sent to a Class 1 disposal site. Instead, after they are completely dry inside, take them to a pesticide container recycling center or a Class 2 disposal site. Check with recycling managers or local disposal site personnel to determine whether you have to have the containers inspected by the county agricultural commissioner's office before you can dispose of them. To prevent any chance that these containers might be used for other purposes before you have a chance to dispose of them properly, cut or punch holes in them to make them unusable.

After adding the pesticide and any additional chemicals needed for the application, fill the spray tank to its final volume. Do not allow the tank to overflow during filling. Also, never let the filling hose, pipe, or other filling device come into contact with liquid in the tank. If you fill the tank through a top opening, leave an air gap between the spray tank and the filling device.

FIGURE 6-13.

Carefully pour pesticides into the spray tank. Rinse measuring containers after use, triple-rinse empty liquid pesticide containers, and pour the rinse solutions into the spray tank.

FIGURE 6-14.

When filling a spray tank, make sure that there is an air gap between the filler pipe and the top level of the water in the tank. This prevents backflow of pesticide-contaminated water into the water supply.

This space should equal at least twice the diameter of the filling pipe to keep the spray mixture from siphoning back into the water supply after you stop the water flow (Figure 6-14). Side- or bottom-filling systems require check valves to prevent backflow of pesticides into the water supply.

See Sidebar 16 for examples of the amount of pesticides that are removed from containers through triple-rinsing.

Service Containers

A service container is any container other than the original labeled packaging that holds pesticides. Manufacturers design these for application, storage, or transportation of pesticide concentrates or use-diluted preparations. Always label service containers with the following information:

- common name of the pesticide
- the signal word *(Danger Poison, Danger, Warning,* or *Caution)*
- name and address of the person responsible for the container

APPLYING PESTICIDES EFFECTIVELY

To use pesticides safely and effectively, you must confine them to the treatment area and apply them in proper amounts. Pesticide coverage should usually be uniform in order to be most effective. For example, to control some plant-feeding pests, the pesticide spray must coat both sides of leaf surfaces.

Spills, leaks, and drift are wasteful of the pesticide and may leave residues in nontarget areas or may contaminate nontarget areas, damaging them or making them unusable for a time. Improper equipment calibration results in too little or too much pesticide reaching the target site. Effective pesticide applications require that you know and follow standards of care, including

- using proper equipment that is in good condition
- developing good application techniques
- reducing or eliminating drift
- having an awareness for all potential hazards

SIDEBAR 16

Triple-Rinsing Procedures for Pesticide Containers

PROCEDURE:

1. When container is empty, let it drain into spray or mixing tank for at least 30 seconds.
2. Add correct amount of water to container as follows.

CONTAINER SIZE	RINSE SOLUTION NEEDED
5 gallons or less	1/4 of the container volume
More than 5 gallons	1/5 of the container volume
28 gallons or more	Does not require triple-rinsing. Return to dealer.

3. Close the container.
4. Shake or roll the container to get solution on all interior surfaces.
5. Drain the container into your sprayer or mixing tank. After the container is empty, let it drain for an additional 30 seconds.
6. Repeat steps 2 through 5 ***two*** additional times.

AMOUNT OF ACTIVE INGREDIENT (A.I.) REMOVED FROM A 5-GALLON CONTAINER BY TRIPLE-RINSING:

	AMOUNT OF ACTIVE INGREDIENT (A.I.) REMAINING*
Drain	14.1875 grams a.i.
1st Rinse	0.2183 gram a.i.
2nd Rinse	0.0034 gram a.i.
3rd Rinse	0.00005 gram a.i

*After draining, a 5 gallon container may contain as much as 1 ounce of formulated pesticide. This would amount to 14.1875 grams of a.i. for a formulation containing 4 pounds of a.i. per gallon.

SELECTING APPLICATION EQUIPMENT

Be sure that the equipment you use to apply pesticides is suited to the location and conditions of the treatment area. Equipment that is too big or powerful may be as much of a problem as equipment that is too small. Most pesticide application equipment works efficiently only in a limited number of situations.

Choose application equipment that is comfortable to work with and easy to use. Make sure the equipment is easy to repair and that parts are readily available. Hand-held equipment must be lightweight so that it will be convenient to use. Motor-powered units should be quiet enough to prevent operator stress, yet powerful enough to do the job properly. Moving parts need shields and guards to prevent accidents and injuries. Powered equipment must have accurate gauges that allow you to monitor spray pressure, tank volume, and other functions.

Make sure that filler covers on spray tanks close properly, seal well, and are lockable (Figure 6-15). Hoses and fittings should be strong and durable to prevent loss of pesticide mixtures and environmental contamination.

FIGURE 6-15.

Make sure the tank cover fits tightly to prevent pesticide mixtures from splashing out during operation or while you are transporting the equipment. If tanks are ever left unattended, their covers must be locked.

Safe Application Techniques

Just as for effective pesticide application, safe application techniques require that you know and follow standards of care, including

- working under proper weather conditions
- controlling droplet size and deposition
- having an awareness of the application site and its hazards
- developing special application patterns for the site to accommodate hazards and environmental conditions
- leaving unsprayed buffer strips to protect people or sensitive areas

Working under Proper Weather Conditions. Weather has an important influence on pesticide applications in outdoor areas. Its effect on pesticide applications in interiorscapes and other confined spaces is more subtle. Temperature affects the phytotoxicity of certain pesticides, so label directions usually warn against using these products when temperatures are above or below critical limits. Warm temperatures associated with clear, sunny weather speed up the breakdown of pesticides. Ultraviolet light, which is most intense during these times, contributes to this breakdown.

Rainfall, fog, and even heavy dew affect pesticide applications. This is because the moisture dilutes pesticides and may wash the material off of treated surfaces. Rainwater washes pesticides into the soil, creating the potential for groundwater and surface water contamination. Water movement after heavy rains carries pesticides away from the application site.

Wind influences pesticide drift and affects volatilization. Strong air movements are responsible for uneven pesticide deposition at the treated site and for off-site drift. Some degree of air movement that is less than 10 miles per hour, however, can assist in getting good coverage over treated surfaces.

Controlling Droplet Size and Deposition. Droplet size is affected by many factors such as nozzle size, style, and condition, as well as spray volume, spray pressure, and weather influences. Most application equipment emits a spray with a wide range of droplet sizes, but the best spray applications result from uniform-sized droplets that are applied evenly to all treated surfaces. Increase the uniformity of spray droplets by selecting nozzles that are designed for the working pressure and volume of your sprayer. Inspect and replace worn or defective nozzles. Check nozzles at least once a year to make sure their output varies by no more than 10% from the manufacturer's rated output. Application speed is critical. Adjust your speed to the type and size of the area being treated. You must travel more slowly, for instance, when spraying large trees and shrubs because the spray droplets must travel farther and the larger total surface area requires more volume to achieve adequate coverage.

Application Pattern

An application pattern is the route you travel with your application equipment or the method you use while applying a pesticide. A proper application pattern should provide an even distribution of pesticide over the treated area. To do this, you must avoid overlaps or gaps. When treating a large, outdoor area, the pesticide application speed usually determines the uniformity of the application pattern. At higher speeds, the equipment bounces more. When establishing an application pattern, consider

- prevailing weather conditions
- what chemical you are spraying
- hazards in or near the application site

Design a pattern that eliminates any need to contact or travel through airborne spray, walk or drive through freshly treated areas, or be exposed to dripping foliage. In interior applications, particularly when treating an atrium or large tree specimens, spray the lower foliage first and then move

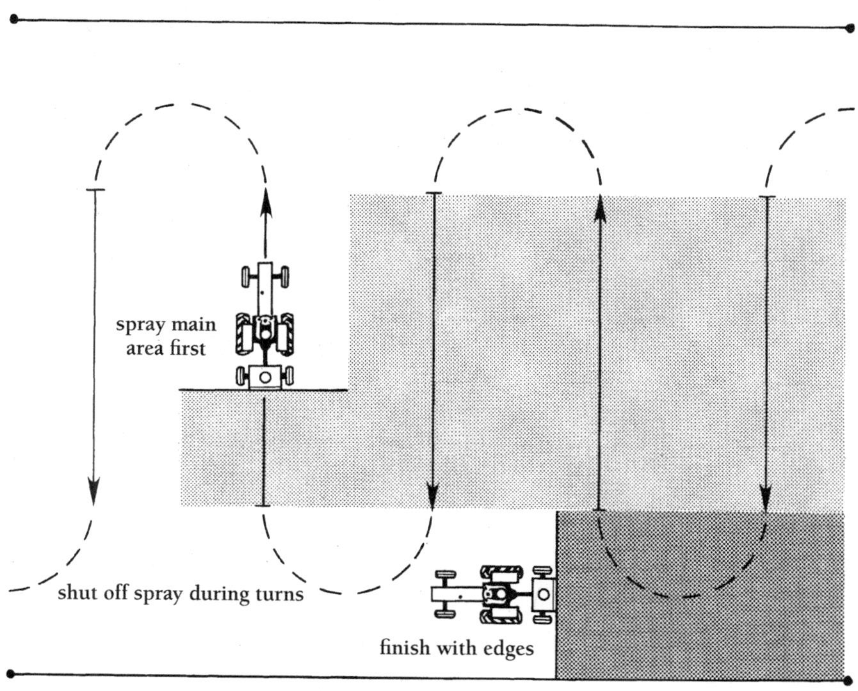

FIGURE 6-16.

Shut off spray nozzles during turns to prevent uneven applications. After spraying the main part of the treatment area, finish by spraying the edges.

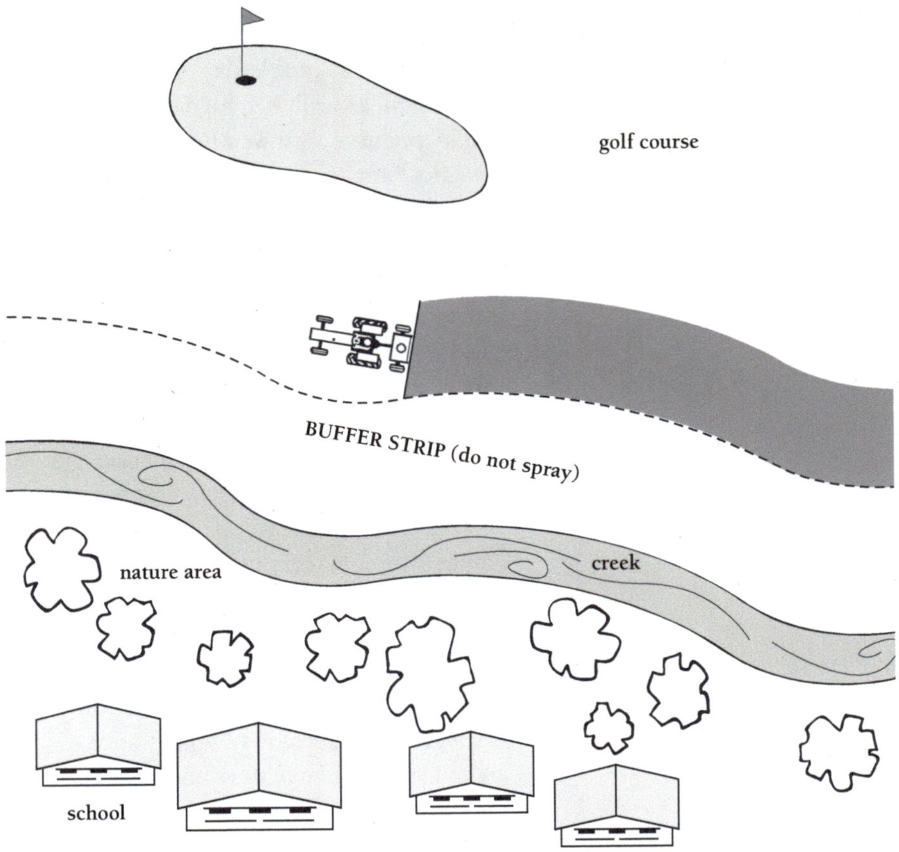

FIGURE 6-17.

Leave untreated buffer strips between the application site and adjoining sensitive areas. The buffer strip should be more than the width of one spray swath.

vertically up the plant. This minimizes pesticide runoff onto the operator. When treating the undersides of leaves with hand-held equipment, never point the nozzles toward yourself.

Operating equipment (such as a boom applicator) produces an uneven application during turns. Figure 6-16 illustrates how to make a uniform application by shutting the sprayer off during turns. Watch for clogged nozzles that will also produce uneven applications.

Leave untreated buffer strips when the area you are treating adjoins sensitive areas. These are locations where applications might expose organisms, people, water bodies, or structures to pesticides (Figure 6-17). Generally, the buffer strip should be no less than the width of one spray swath. The size of the buffer strip depends on

- the type of application equipment you are using
- prevailing weather conditions, especially direction and speed of air movement
- the nature of the pesticide being applied
- the type of pest problem being treated
- the sensitive nature of adjoining areas

Pesticide Drift

Pesticide drift refers to the movement of pesticides away from the treatment site. Drift is most serious when you make applications under windy conditions, especially when using high pressure and small-orifice nozzles. Wind can move spray droplets away from the intended treatment area (Figure 6-18). Another form of drift occurs when

FIGURE 6-18.

Pesticide droplets will be more likely to drift when you make applications if there is strong air movement. Chances of drift are increased by small droplets that result from high spray pressure and small nozzle orifice size.

sprayed pesticides partially evaporate before reaching their target. The resulting vapor can drift away from the treatment area through air movement, often traveling several miles.

Sidebar 17 lists some of the factors that influence drift. There are many steps you can take to reduce drift. Elimination of very small droplets has a significant effect. Increase the droplet size by using larger nozzles and by lowering the output pressure of the sprayer. Certain adjuvants, called deposition aids, assist in increasing droplet size or reducing the potential for evaporation. Add one of these materials to the spray tank to help prevent drift.

Do not spray during windy conditions. Usually, winds of less than 5 miles per hour help provide good pesticide distribution in trees and leafy plants. Stronger winds, however, increase the potential for drift. In some cases, spraying may be illegal if the wind speed exceeds a designated rate. Contact the county agricultural commissioner for information on pesticide application restrictions during windy conditions.

Some conditions, such as a temperature inversion, increase the adverse effects from drift of small droplets and vapors. Other conditions with high temperatures and low humidity increase the evaporation rate, and this reduces the size of droplets before they reach their target. The resulting smaller droplets may be highly subject to drift.

SIDEBAR 17

Factors that Influence Pesticide Drift

PESTICIDES:

- ☐ Volatility of the active ingredient
- ☐ Solvent used to dissolve or suspend the active ingredient (formulation)
- ☐ Diluent used to dilute or mix the pesticide in the spray tank

ADJUVANTS:

- ☐ Presence of deposition aids, thickeners, and stickers (These reduce drift by making droplets larger or less volatile.)

APPLICATION EQUIPMENT:

- ☐ Operating pressure of the spraying system
- ☐ Nozzle size
- ☐ Distance from nozzles to the target surface
- ☐ Height from which the spray is released
- ☐ Travel speed of the application equipment
- ☐ Application pattern and technique

TARGET SURFACES:

- ☐ Size of the target area
- ☐ Location of the target area
- ☐ Nature of the target surfaces

WEATHER CONDITIONS:

- ☐ Wind intensity
- ☐ Wind direction
- ☐ Air temperature
- ☐ Humidity

CLEANUP AND DISPOSAL

Disposal of Surplus Diluted Pesticide

To avoid problems associated with leftover pesticide mixtures, calculate the exact size of the treatment area and mix only enough pesticide for the job. If you have some spray mixture left over, use it in another use site listed on the label if you can. Otherwise, you must send the leftover material to a Class 1 disposal site. There are private companies that specialize in collecting and transporting pesticide wastes to Class 1 disposal sites.

Never indiscriminately dump excess pesticide. It is illegal and is a potential source of environmental and groundwater contamination. If you are convicted of dumping you will be subject to a large fine and possibly a jail term.

Pesticide Container Disposal

Regulations concerning the disposal of pesticide containers vary from county to county. Obtain specific disposal information from the local Water Quality Control Board, the Department of Health Services, and the county agricultural commissioner. This is the recommended action to take to find out how to properly and legally dispose of paper pesticide bags.

Through pesticide container recycling centers, you can recycle metal and plastic containers if they have been properly triple-rinsed and are dry inside. Contact the county agricultural commissioner for disposal requirements and the location of a pesticide container recycling center in your area. As an alternative to recycling, you can take rinsed containers to an approved Class 2 disposal site. Any unrinsed or partially filled containers must be transported to a Class 1 disposal site. Check with the Water Quality Control Board or county agricultural commissioner in your area for the locations of Class 1 and Class 2 disposal sites and for requirements regarding transport of these materials. Containers may need to be inspected by a county agricultural commissioner staff member before being transported to the disposal site.

Cleaning Application Equipment

After each use, you must clean and decontaminate the application equipment. Otherwise, residues remaining in tanks may contaminate a subsequent pesticide mixture and could alter its toxicity or damage the plants being sprayed. Pesticide residues on the outside of application equipment can be hazardous to the people who operate or repair the equipment. Wash the outside of spray equipment with water, using a small amount of detergent if necessary. Clean the equipment right at the application site whenever possible. If cleaning at the application site is not possible, you will have to contain the runoff. You may dispose of this runoff water at the application site or, if you will be making further applications of the same pesticide in the immediate future, put it into the spray tank for use with subsequent applications. Otherwise, you must have this rinse water transported to a Class 1 disposal site.

Rinse the inside of the tank with water. If necessary, decontaminate it using an appropriate pesticide tank cleaning material. You can buy commercial pesticide tank cleaning and neutralizing compounds from chemical suppliers and farm equipment dealers. Be sure to check the pesticide label for any precautions regarding the use and disposal of cleaning and decontaminating chemicals. Follow the directions for the cleaner you use for your spray tank. Be sure to run pumps and agitators and flush all hoses.

Personal Cleanup

After using pesticides, you must clean and store your PPE, bathe thoroughly, and change into clean, uncontaminated clothing. Never eat, drink, smoke, or use the bathroom until you have thoroughly washed. When bathing, take special care to wash your hair and clean under your fingernails.

Place work clothing that you wore during the pesticide application into a plastic bag until you can launder it. Wash work clothing separately from other clothing. After using a washing machine to launder this work clothing, run a full cycle on the washer, with soap but without clothes, to remove any possible residues (see Sidebar 14).

RECORDKEEPING

Maintain records of every pesticide application you make and other activities associated with your use of pesticides. Keep records of the following pertinent information:

- The pesticide application:
  - name, manufacturer, and EPA registration number of the pesticide
  - total amount of pesticide used
  - amount of water used
  - calibration adjustments
  - adjuvants added to the mixture
  - type of equipment used or method of application
  - severity of pest infestation
  - restricted-entry interval
  - date and time when the application was completed
- Other records:
  - pesticides handled by employees
  - handler training specific to the classes of pesticides handled by employees

Also, note the temperature and general weather information at the time of application. Write down a note of any other conditions that might have an influence on the effectiveness of the pesticide. Keep a record of the names of people you spoke to regarding each pesticide application. Include any follow-up information and notes of application results. If the application was recommended by a pest control adviser (PCA), keep a copy of the PCA's written recommendation with your application records.

Application records are helpful as a reference for your history of pesticide use and are vital in case problems associated with the application should develop. Good records may also be important to your defense in any legal action.

LIABILITY

You assume personal responsibility for accidents and injuries that might arise from each pesticide application. You may be subject to fines, jail time, and loss of your applicator certificate or license if you are negligent in your application of pesticides or if you break state or federal laws. In addition, courts may hold you responsible in lawsuits for personal injury or damages. If you are working for someone else, your actions may result in lawsuits and fines against your employer as well. Should someone bring a claim of negligence against you, accurate records of all your pesticide applications help in your defense.

Should the pesticide you are applying drift and damage nearby plants or animals or someone's belongings, or should the drift cause human injury, you may incur personal liability. The pesticides you apply can even cause inadvertent damage to the plants or surfaces that you intended to spray. Damage may result from

- improper mixing
- using the wrong adjuvants
- using an improper application method
- applying the wrong pesticide
- using a pesticide that has been contaminated with impurities (such as other pesticides if the tank has not been properly cleaned)
- using pesticides under inappropriate environmental conditions

In addition to the hazards described

above, pesticides and pesticide application equipment are attractive nuisances. Children, fascinated with what you are doing, may be injured by chemicals and equipment if you leave these unattended.

Often, taking prompt action will greatly reduce the extent of damage (and therefore liability) from a pesticide accident or application error. See the next chapter for information on how to deal with pesticide emergencies.

Liability Insurance and Workers' Compensation Insurance

Commercial pesticide applicators and maintenance gardeners must have liability insurance to protect themselves from claims associated with pesticide use. Clients of professional applicators may require that the applicators have liability insurance. "Errors and omissions" policies cover the costs of damage resulting from accidents and improper use of pesticides. Insurance companies consider an applicator to be a better risk if he or she maintains complete records and is conscientious in efforts to use pesticides responsibly.

Some professional organizations representing pesticide applicators have information on insurance companies and policies. Some provide opportunities for members to participate in group policies. Select insurance that is suitable to your operation and specialty. Be sure you understand the extent of coverage and policy liability limits.

California state laws also require that employers carry workers' compensation insurance for all employees. An application for the Pest Control Business License requires that you include the workers' compensation insurance carrier's name and the policy number and expiration date of your policy. See Chapter 4, Pesticide and Pest Control Laws and Regulations, for financial responsibility requirements.

CHAPTER 6: REVIEW QUESTIONS

1. The best way to avoid most pesticide accidents is by

☐ a. recognizing common symptoms of pesticide poisoning
☐ b. understanding the pesticide chemical characteristics
☐ c. keeping up to date on pesticide technology
☐ d. following the pesticide label and obeying the laws and regulations that deal with pesticides

2. The mandatory training for employees who handle pesticides as part of their work does NOT include how to

☐ a. handle, open, and lift containers
☐ b. confine spray to the target area
☐ c. recognize the chemical characteristics of the pesticide
☐ d. recognize pesticide poisoning symptoms

3. Why is training on recognizing and avoiding heat stress important for pesticide handlers?

☐ a. Many heat stress symptoms are similar to pesticide poisoning symptoms
☐ b. Pesticide poisoning is usually caused by heat stress
☐ c. Heat stress is usually caused by pesticide poisoning
☐ d. Prevention of heat stress requires the use of personal protective equipment

4. Who is responsible for providing personal protective equipment to pesticide handlers?

☐ a. the University of California
☐ b. county agricultural commissioners
☐ c. employees are required to provide their own equipment
☐ d. the employer of the pesticide handler

5. On a Material Safety Data Sheet (MSDS), you will NOT find information on

☐ a. fire and explosion hazards of the pesticide
☐ b. emergency spill or leak cleanup procedures
☐ c. application use directions or requirements
☐ d. health hazards of the pesticide

6. What criteria must you use to select the personal protective equipment for applying a specific pesticide?

☐ a. written pesticide use recommendation
☐ b. pesticide label and regulation requirements
☐ c. pest management guidelines
☐ d. the Material Safety Data Sheet (MSDS) requirements

7. An advantage of wearing coveralls over your clothing while applying pesticides is that they

☐ a. can easily be removed if they become contaminated
☐ b. keep you cooler
☐ c. protect you from heat stress
☐ d. are required for most applications

8. Leather and fabric gloves are unsuitable for pesticide handling because they

☐ a. are very expensive
☐ b. absorb pesticides
☐ c. cause uncomfortable sweating of the hands
☐ d. break down when exposed to pesticides

9. Employees handling pesticides must wear eye protection if they are

☐ a. working inside an enclosed cab
☐ b. applying a solid bait form of a rodenticide
☐ c. using a boom sprayer with the nozzles behind you and pointing down
☐ d. spraying with a backpack sprayer

10. While applying certain pesticides, you must select and use appropriate respiratory protection based on

☐ a. county agricultural commissioner requirements
☐ b. physician recommendations
☐ c. current weather conditions
☐ d. pesticide label requirements

11. Cleaning of personal protective equipment, including body wear, should be done

☐ a. after using pesticides
☐ b. weekly, if the equipment is used for more than two days
☐ c. weekly, if the equipment has visible residues on it
☐ d. before a different pesticide is handled

12. When transporting pesticides in a vehicle,

☐ a. secure the packages in the passenger area
☐ b. keep containers as close as possible to the driver
☐ c. secure containers in the cargo area
☐ d. keep containers as far away from the driver as possible

13. An important reason to inspect the area where you will be applying pesticides is to

☐ a. estimate how long the application will take
☐ b. become aware of conditions or objects that may affect the safety of the application
☐ c. schedule the application with other applications in the area
☐ d. locate a mixing site

14. The Physical Properties section of a MSDS provides information about a pesticide's

☐ a. effect on human health
☐ b. unusual fire or explosion hazards
☐ c. boiling, melting, and freezing points
☐ d. chemical family and name

15. What is the maximum distance from the application or mixing site that soap, single-use towels, and water for emergency washing can be located?

☐ a. 25 feet
☐ b. 100 feet
☐ c. 500 feet
☐ d. ¼ mile

16. What are the three categories of training that must be covered when training pesticide handlers?

☐ a. mixing techniques, application techniques, and cleanup methods
☐ b. using pesticides safely, emergencies and health, and general information
☐ c. handling containers, storing pesticides, and disposal
☐ d. maintaining PPE, environmental protection, and poisoning symptoms

17. Employers of pesticide handlers are responsible for

☐ a. providing and maintaining all label- and regulation-required PPE
☐ b. providing PPE but not maintaining it
☐ c. maintaining PPE but not providing it
☐ d. providing and maintaining only the minimal PPE required by California regulation

7 Pesticide Emergencies

FIGURE 7-1.

Any time you work around pesticides, you should be prepared to handle an emergency. A pesticide emergency may be the result of leaks, spills, fires, thefts, misapplication, or improper storage or handling.

OCCASIONALLY, an accident or other problem will inevitably occur while you are handling or applying pesticides, even if you are working under the most careful of conditions. Many pesticides are still hazardous when diluted with water, but undiluted pesticides are usually much more dangerous (Figure 7-1). Pesticide emergencies may result from such incidents as

- leaks
- spills
- misapplication
- lack of care during storage or handling
- failing to use label- or regulation-required personal protective equipment (PPE)

Be especially aware of your responsibilities as an employer regarding arrangements for emergency medical care and provision of certain information to employees who handle pesticides. Also, provide employees with basic decontamination supplies. Whenever you or your employees use pesticides, have the names and locations of nearby medical facilities that are capable of treating pesticide-related injuries available at the application site or in your vehicle. If a leak or spill exposes you or an employee to the pesticide, seek immediate medical care. Have someone drive the exposed person to the care facility.

Be prepared to offer first aid to anyone who is exposed to pesticides. After administering first aid, insist that they receive prompt medical attention. To help employees locate medical facilities, post information about medical facility locations in a conspicuous place at the worksite or in service vehicles.

FIRST AID

First aid is the help that you give a person who has been exposed to pesticides before he or she receives professional medical help. First aid is not a substitute for professional medical attention. The *precautionary statement* section of each pesticide label provides specific first aid information.

Injury can occur if a pesticide splashes onto your skin or into your eyes or is accidentally swallowed, or if you inhale its vapors, dusts, or fumes. The type and amount of exposure determines what first aid and medical treatments are required. Serious pesticide poisoning may cause a person to stop breathing or may cause convulsions, paralysis, skin burns, or blindness. Timely first aid treatment for pesticide exposure may reduce the extent of injury and even save a life. A helpful way to prepare yourself for such emergencies is to complete an American Red Cross first aid course that includes cardiopulmonary resuscitation (CPR) training.

Protect yourself and others whenever you administer first aid to a person suffering from pesticide exposure. Avoid getting pesticides onto your skin and do not inhale vapors. Do not enter an enclosed area to rescue a person who has been overcome by toxic pesticide fumes unless you have the proper self-contained or supplied-air respiratory equipment. Remember, the pesticide that affected the injured person can also injure you.

Get professional emergency medical care *at once* for anyone who has been exposed to a highly toxic pesticide or who shows signs of pesticide poisoning. Call an ambulance (call 9-1-1) or transport the injured person to a medical facility for treatment. Provide medical personnel with complete information about the pesticide suspected of causing the injury. In addition to the first aid measures listed below, the speed with which you obtaining medical care can often determine the extent of injury.

Pesticides on Your Skin or Clothing

Concentrated pesticides spilled onto your skin or clothing can cause serious injury (Figure 7-2). Some pesticides may cause skin burns or rashes or may even, through skin absorption, produce internal poisoning. Immediately remove contaminated clothing and wash the affected areas of skin with clean water and soap. Do not reuse the contaminated clothing or equipment until it has been decontaminated. If these items cannot be adequately decontaminated, you must properly dispose of them.

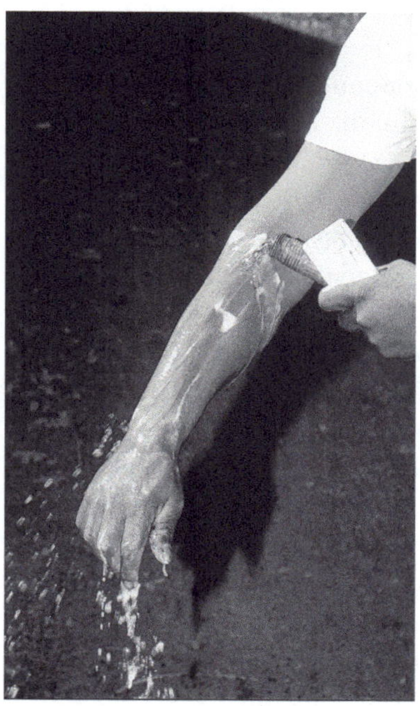

FIGURE 7-2.

If pesticides spill on the skin, the first step is to remove contaminated clothing and wash the affected body parts with soap and plenty of water. Do this quickly to avoid serious injury.

First Aid for Skin Exposure

Follow these steps in case you, an employee, or anyone else receives skin exposure to pesticides:

- **Leave the Contaminated Area.** Get away (or remove the victim) from the fumes, spilled pesticide, and further contamination. Do this quickly!
- **Prevent Further Exposure.** Remove the contaminated clothing and thoroughly wash the affected skin and hair areas. Use soap or detergent and large amounts of clean water.
- **Get Medical Attention.** Call an ambulance or have someone transport the injured person to the nearest medical facility as quickly as possible.

Pesticides in the Eyes

Many pesticides are caustic and cause serious damage if they get into your eyes. Prompt first aid, followed by medical care, helps limit the extent of the damage.

First Aid for Eye Exposure

Follow these steps in case you, an employee, or anyone else receives eye exposure to pesticides:

- **Leave the Contaminated Area and Remove the Exposed Person from the Source of Contamination.**
- **Flush the Eyes.** Immediately flush the affected eye or eyes with a gentle stream of clean, running water of a temperature that will not cause further injury (not too hot or too cold). Hold the eyelids open to assure thorough flushing (Figure 7-3), directing the water onto the bridge of the affected person's nose, rather than directly into the eyes. When flushing the eyes, do not add any chemicals or drugs to the water: that could increase the extent of injury. If no running water is available, slowly pour clean water from a glass, water cooler, or other container.
- **Obtain Medical Care.** Always get professional medical attention after the flushing. Let the medical providers know the name of the pesticide that was involved in the exposure.

Inhaled Pesticides

Inhaled chemicals such as fumigants, pesticide dusts, vapors from spilled pesticides, and fumes from burning pesticides can cause serious lung injury and may be absorbed into other parts of the body through the lungs. Perform first aid measures immediately to reduce the extent of injury and prevent death.

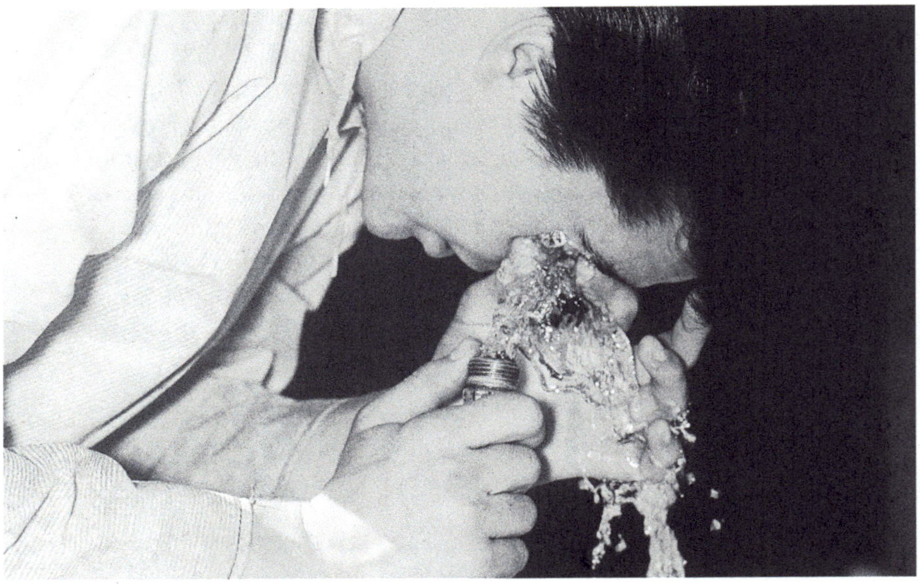

FIGURE 7-3.

If a pesticide gets into the eyes, flush them thoroughly with clean, running water. Then get medical attention.

Wear a self-contained or supplied-air respirator when entering an enclosed area to rescue a person who has been overcome by pesticide fumes. Cartridge respirators are not suitable for high concentrations of pesticide vapors or deficient-oxygen conditions. If you do not have a supplied-air respirator, call for emergency help. You will be doing more to help the injured person if you seek proper emergency help than if you go in on your own and are overcome by the pesticide fumes yourself.

First Aid for Pesticide Inhalation

Follow these steps in case you, an employee, or anyone else inhales pesticide vapors:

- **Leave the Contaminated Area and Remove the Exposed Person from the Contaminated Area.** Anyone who has been overcome by pesticide vapors must get to fresh air immediately. Avoid subjecting the victim to physical exertion, since this places an extra strain on the heart and lungs.
- **Loosen Victim's Clothing.** Looser clothing makes breathing easier and releases pesticide vapors trapped between the clothing and the skin.
- **Restore Breathing.** If the victim's breathing has stopped or is irregular or labored, begin artificial respiration (rescue breathing). Continue with rescue breathing until the victim's breathing has improved or until medical help arrives. If the victim has stopped breathing and has no pulse, begin cardiopulmonary resuscitation (CPR) and continue until medical help arrives.
- **Treat for Shock.** Inhalation injury often causes a person to go into shock. Keep the injured person calm and lying down. Prevent chilling by wrapping the person in a blanket after you remove any contaminated clothing. Do not give the victim any alcoholic beverages.
- **Watch for Convulsions.** If convulsions occur, protect the victim from falls or injury and keep air passages clear by making sure the victim's head is tilted back.
- **Get Immediate Medical Care.** Call an ambulance or transport the person to the nearest medical

facility. Let the medical providers know the name of the pesticide that was involved in the exposure.

Swallowed Pesticides

Two immediate dangers are associated with swallowing pesticides. The first relates to the pesticide's toxicity and the poisoning effect it will have on a person's nervous system or other internal organs. The second involves any physical injury that the swallowed pesticide may cause to the linings of the mouth and throat and to the lungs. Corrosive materials (those that are strongly acidic or alkaline) can seriously burn these sensitive tissues. Petroleum-based pesticides can cause lung and respiratory system damage, especially during vomiting. Never induce vomiting without first checking the label for specific directions and any warnings against inducing vomiting.

You can reach regional California Poison Control System (CPCS) centers, located in Sacramento, San Francisco, Fresno, and San Diego, by telephone at any time (Figure 7-4). In a poisoning emergency, call CPCS from anywhere in California by using a single toll-free number: 1-800-222-1222. CPCS centers provide quick, lifesaving information on treatment for poison victims. Contact 9-1-1 emergency if the poison victim cannot wake up, has difficulty breathing, or has convulsions. For detailed information from CPCS, visit their Web site at http://www.calpoison.org.

First Aid for Swallowed Pesticides

Act quickly when someone swallows a pesticide. Follow the pesticide label or California Poison Control System instructions or take the following steps if an employee or anyone else swallows a pesticide:

- **Leave the Contaminated Area and Remove the Exposed Person from the Contaminated Area.**
- **Dilute the Swallowed Pesticide.** If the person is conscious and alert, get him or her to drink large amounts (1 quart for an adult or a large glass for a child under 7) of water or milk. Do not give any liquids to a person who is unconscious or convulsing.
- **Induce Vomiting.** If the label

Pesticide Poisoning Emergencies

California Poison Control System

Managed by University of California, San Francisco, School of Pharmacy, the California Poison Control System consists of four answering sites:

- UC Davis Medical Center, Sacramento
- San Francisco General Hospital, San Francisco
- Valley Children's Hospital, Fresno
- UC San Diego Medical Center, San Diego

In a poisoning emergency, the Poison Control System can be reached from throughout California by using a single toll-free number:

1-800-876-4766 (1-800-8-POISON)

TTY 1-800-972-3323

Health education services are available throughout the state. For information regarding pamphlets, stickers, or presentations call:1-800-582-3387.

FIGURE 7-4.

You can reach the California Poison Control System (CPCS) by telephone at any time throughout California from a single toll-free phone number. This center provides quick, lifesaving information on poisoning treatment. A national poison control number 1-800-222-1222 is also available.

directs you to, induce vomiting. Make sure that the victim is kneeling or lying face-down or on his or her right side. *If in doubt, do not induce vomiting.*

- **Obtain Medical Care.** Call an ambulance or transport the poison victim to the nearest medical facility. Provide medical personnel with the name of the swallowed pesticide and as much other information as possible about the pesticide.

PESTICIDE LEAKS AND SPILLS

Treat any pesticide leak or spill as an emergency. Concentrated pesticide spills are much more dangerous than spills of pesticides that have been diluted with water, but you should deal with both types immediately. Leaks or spills can occur during transport, storage, or use of pesticides.

When a spill occurs on a public roadway, immediately contact the nearest California Highway Patrol Office and the State of California Office of Emergency Services (1-800-852-7550). For non-emergency calls and general information, you can contact the California Highway Patrol at 1-800-TELL CHP (1-800-835-5247). These agencies will take charge of coordinating the cleanup and protecting the public. In the event of a pesticide spill on a public roadway, you are required to file a report with the California Office of Emergency Services. If leaks or spills should occur in areas other than public roadways, follow the emergency procedures listed below. Report all leaks or spills of pesticides, no matter where they occur, to the local county agricultural commissioner as soon as possible.

When you clean up a spill, put the spilled material and anything that was contaminated by the spill into a sealable drum. Seal the drum and label it to indicate that it contains hazardous waste. The label should include the name of the pesticide as well as its signal word (*Danger Poison, Danger, Warning,* or *Caution*).

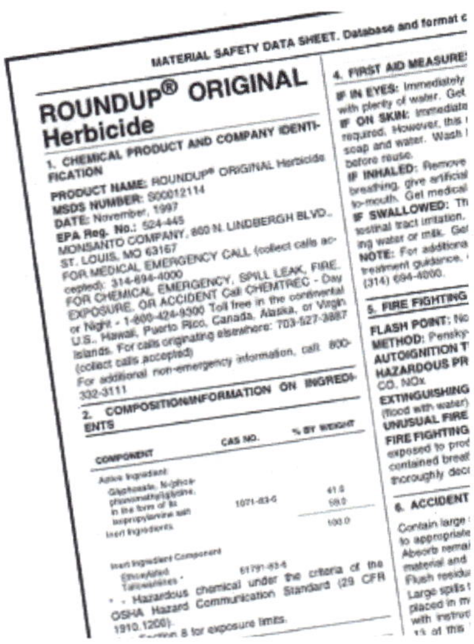

FIGURE 7-5.

The MSDS provides helpful information and precautions on cleaning up a pesticide spill.

Because local regulations vary, contact the local county agricultural commissioner or Water Quality Control Board for instructions on how to dispose of the sealed drum and its contents. Under most circumstances, you will have to send the residue from a pesticide spill to a Class 1 disposal facility.

Spills on cleanable surfaces, such as concrete, require thorough decontamination. Commercial decontamination products are available for this purpose; alternatively, you can prepare a solution dissolving 4 tablespoons of detergent and 1 pound of soda ash (sodium carbonate) per gallon of water. Soda ash cannot be used for detoxification in the case of a few pesticides, so check the pesticide label or the "Accidental Release" section of the MSDS (Figure 7-5) before you use this solution. Contact the pesticide manufacturer if you have any questions. Soda ash (sodium carbonate) is available from fertilizer dealers.

If anyone has been injured or contaminated in the spill, administer first aid. Send for medical help if necessary or have the injured person transported to receive emergency medical help.

Barricade. Rope the area off, set up barricades, or otherwise keep people away from the contaminated site.

Ventilate the Area. If the spill is indoors, get people out of the building. Open doors and windows. Set up a portable fan to exhaust the building's air to the outside and away from any area where people are or may occupy or pass through.

Cleaning up Pesticide Leaks or Spills

The clean-up of major pesticide spills requires the help of professionals. It is extremely difficult and costly to remove contaminated soil or to prevent or clean up groundwater contamination. The types of pesticide leaks and spills you are most likely to encounter involve controllable quantities, such as when a container is damaged or slips to the ground or when a diluted pesticide leaks from application equipment. You must take proper and immediate response to even these types of small leaks and spills in order to minimize damage to human and environmental health.

There are certain basic steps you should follow in cleaning up a pesticide leak or spill:

- **Clear the Area.** Keep people and animals away from the contaminated area. Provide first aid to anyone who has been injured or contaminated. Send for medical help if necessary. Do not allow any smoking near a spill. If the spill occurred in an enclosed area, shut off all electrical appliances and motors that could produce a spark and ignite combustible fumes to start a fire or explosion.
- **Wear Personal Protective Equipment.** Before you begin any cleanup, put on the personal protective equipment listed on the pesticide label for handling the concentrated material. Check the pesticide label for additional precautions, but when you are uncertain as to just what has been spilled, wear the maximum protection. This includes chemical-resistant boots and gloves, waterproof protective clothing, goggles, and a respirator.
- **Contain the Leak.** Stop the leak by transferring the pesticide to another container or by patching the leaking container (repair paper bags and cardboard boxes with strong tape). Use soil, sand, sawdust, or absorbent clay to form a containment "dam" around liquid leaks. Common cat litter is a good absorbent material for use in pesticide cleanup. If the wind is blowing pesticide dusts or powders, spray the area lightly with water or cover the spill with a plastic tarp to prevent off-site movement.
- **Clean up the Pesticide.** Proceed to clean up the spill or leak (Figure 7-6). In the case of a liquid spill, brush the containment dam of absorbent material toward the center of the spill. Add additional absorbent material if necessary. If the spill is on soil, shovel out any contaminated soil for disposal. For large spills, you may need to have a hazardous waste cleanup company run tests to determine how much soil to remove. For small spills, remove soil to an inch or two beyond the moistened area. Place the absorbent or spilled dry product and any contaminated soil in a sealable container. Containers for holding contaminated materials must be suitable for transportation and must be sealable. Seal the container and label it with the pesticide's name and signal word.
- **Clean Nonporous Surfaces and Safety Equipment.** If the spill occurred on a cleanable surface such as concrete or asphalt, use a broom to scrub the contaminated surface with a strong detergent solution. Then clean it up again with absorbent material and place it in the sealable disposal container. Clean or dispose of equipment that is contaminated during cleanup, such as brooms, shovels, and dustpans. Clean your personal protective equipment when you finish.
- **Dispose of the Material.** Local regulations on disposal of hazardous materials may vary. Check with your local county agricultural commissioner or Water Quality Control Board for instructions on how to dispose of the sealed container and its contents.

FIGURE 7-6.

Cover pesticide spills with an absorbent material and shovel this material into a sealable container. When the cleanup is completed, seal and label the container and send it to a Class 1 disposal site. Wear the label-required mixer/loader PPE during cleanup.

MISAPPLICATION OF PESTICIDES

Another form of emergency may exist when a pesticide has been misapplied.

- *Intentional misapplication* involves the intentional use of a pesticide on an unregistered site or the knowing application of pesticides in a manner inconsistent with label directions or regulation requirements.
- *Accidental misapplication* involves the unknowing application of a pesticide to a site not listed on the label.
- *Negligent application* involves the improper calibration of application equipment as well as the improper use and disposal of the pesticide; it also involves the application of pesticides at the wrong time or in any other way that is inconsistent with label requirements

An application mistake is a serious problem; do not compound the damage by failing to take responsible, corrective action once you discover the mistake. You may be financially responsible for damages, both physical and legal, caused by your misapplication of a pesticide, but you may be able to reduce the amount of damage and liability by taking prompt action. Of primary importance is the protection of people, animals, and the environment.

Incorrect Amount of Pesticide Used

The application of an insufficient quantity of a pesticide usually fails to provide adequate control of the target pest and can waste time and money, but in general it presents no immediate problems to people or the environment. The application of excessive amounts of pesticide, however, is illegal and can constitute an environmental threat as well as a danger to human health. In addition, overapplication can cause plant damage. This type of problem can result from

- intentional disregard for the use directions
- poor calibration of your application equipment
- pesticide residues from another

application that are left in the spray tank or hoses
- faulty mixing of chemicals in the spray tank
- misunderstanding of the label statement regarding application rates

Residues from an excessive application of a pesticide may last longer than you expect or an overly concentrated application may cause damage to the treated area.

Correcting the Problem. Once you discover that you have made an improper application, take immediate action. Notify the county agricultural commissioner of the problem and seek information and advice on what remedial actions to take. Contact the pesticide manufacturer to find out what corrective measures they suggest. Remember, speed is of the utmost importance when trying to reduce damage from these chemicals.

Applying the Wrong Pesticide

A lack of attention to your mixing operation or the wrong instructions given to an employee may result in the wrong pesticide being applied to the wrong place. Besides the potential for damage to plants or surfaces in the treatment area, use of the wrong pesticide exposes you and your workers to unanticipated hazards that cannot be mitigated. Mixing and application might take place without the required personal protective equipment and might result in injury to the application personnel.

Correcting the Problem. When you discover that you have mixed or applied the wrong pesticide, contact the county agricultural commissioner for help and then call the pesticide manufacturer. Notify people in the application area, and keep them away until the problem is resolved.

Pesticides Applied to the Wrong Site

Another form of accident involves the application of pesticides to the wrong site. This can be a serious problem if the pesticide label does not list that type of site or if the site is different in any of a number of aspects from the one intended for treatment.

Correcting the Problem. Contact the local county agricultural commissioner and the pesticide manufacturer for assistance. Keep people and animals out of the treated area until it is safe to return.

CHAPTER 7: REVIEW QUESTIONS

1. The name, address, and location of a medical facility capable of treating pesticide-related injuries should be available at the

☐ a. facility office
☐ b. pesticide mixing site, application vehicles, or whenever pesticides are used
☐ c. county agricultural commissioner's office
☐ d. pesticide dealer's facility

2. First aid and other emergency information for pesticide accidents is found

☐ a. in the front section of the local telephone directory
☐ b. on the second page of the pesticide manufacturer's informational brochure
☐ c. on the bottom of the pesticide container
☐ d. in the precautionary statements section of the pesticide label

3. First aid for pesticide spills onto the skin includes

☐ a. removing contaminated clothing and washing with soap and water
☐ b. changing out of the contaminated clothing at the next break
☐ c. removing contaminated clothing, then resting in the shade for 20 minutes
☐ d. going to a medical facility for assistance in washing the affected areas of skin

4. If pesticides get into your eyes, you should always

☐ a. keep your eyes covered with a damp cloth until irritation stops
☐ b. gently blot your eyes thoroughly with a damp cloth
☐ c. flush your eyes thoroughly with clean water
☐ d. flush your eyes thoroughly with clean water and seek medical attention

5. If pesticide vapors are inhaled, first aid procedures include getting to fresh air and

☐ a. having the victim recline and breathe rapidly for 15 minutes
☐ b. sending the victim home to rest for the remainder of the work day
☐ c. loosening clothing, restoring breathing if necessary, and seeking medical care
☐ d. having the victim drink a large quantity of beverage and induce vomiting

6. When someone swallows a pesticide you should

☐ a. have the victim rest in the shade for at least one hour
☐ b. if they are conscious and alert, give large amounts of water or milk to dilute
☐ c. have the victim drink fruit juice and induce vomiting
☐ d. watch for signs of illness and seek medical care if symptoms appear

7. When cleaning up a spilled pesticide, the absorbent and all contaminated materials must be put into sealed containers and

☐ a. buried at least 2 feet deep
☐ b. taken to a Class 2 disposal site
☐ c. shipped to a Class 1 disposal site
☐ d. sent to the local agricultural commissioner's office for disposal

8. If the wrong pesticide has accidentally been applied to a site, you should notify the

☐ a. county agricultural commissioner
☐ b. Water Quality Control Board
☐ c. California Poison Control System
☐ d. California Office of Emergency Services

9. What precaution should you take when administering first aid to a pesticide exposure victim?

☐ a. Do not let others assist you
☐ b. Avoid using water to decontaminate the victim
☐ c. Follow the MSDS decontamination procedures
☐ d. Protect yourself and others from exposure

10. Inhalation of pesticide vapors may often cause a person to

☐ a. sweat profusely
☐ b. go into shock
☐ c. complain of muscle aches
☐ d. develop a serious skin rash

11. If a person has swallowed an unknown pesticide, you should induce vomiting

☐ a. immediately
☐ b. only if the labels tells you to
☐ c. if the victim is conscious
☐ d. under no circumstances

12. Leaks and spills of any pesticide should be reported to the

- ☐ a. California Department of Pesticide Regulation
- ☐ b. county agricultural commissioner
- ☐ c. California Office of Health Hazard Assessment
- ☐ d. county health official

13. An accidental misapplication involves

- ☐ a. intentional use of a pesticide on an unregistered site
- ☐ b. improper calibration of application equipment
- ☐ c. unknowing application of a pesticide to a site not on the label
- ☐ d. use of the wrong amount of pesticide

8 Using Pesticides Effectively

TO ACHIEVE OPTIMUM pest management results, you must use pesticides effectively as well as safely. The benefits of a pesticide application should be at least equal in value to any financial or labor investments involved. Benefits include reduction of health hazards to people and improvement of the health, appearance, and growth of turf and ornamental plants. In this chapter we discuss several ways to improve the effectiveness of pesticide applications.

MAKING PESTICIDE-USE DECISIONS

When pests are damaging plants or interfering with the use, maintenance, or safety of landscape, turf, or interiorscape areas, some form of pest management action is needed. Several management methods will be available to you in most cases, so your first step will be to determine which one is best suited to the circumstances.

Factors Influencing Pesticide-Use Decisions

Pesticide use is one of the options you can select for managing a landscape, turf, or interiorscape pest. This option usually is chosen based on a number of criteria, including

- availability of a pesticide product that is registered for the intended use
- safety for the public and for landscape maintenance workers, as well as for pesticide handlers
- potential for environmental contamination
- effectiveness in comparison to other control methods
- cost in comparison to other control methods
- capacity to achieve speedy control

SELECTING THE RIGHT PESTICIDE

The process of choosing the right pesticide can be complicated. Sometimes you will have several pesticide products to choose from for control of the same pest in a particular situation. To get information about pesticide products for specific uses, consult

- pesticide label books (available in writing from the manufacturers or online at the California DPR Web site [http://www.cdpr.ca.gov])
- UC horticulture advisors and agricultural commissioners
- licensed pest control advisers
- pesticide chemical handbooks
- University of California publications, treatment guides, and pest management guidelines (see Sidebar 18)
- California Department of Pesticide Regulation Web site (http://www.cdpr.ca.gov)

The University of California Statewide Integrated Pest Management Program's *Pest Notes,* which include pesticide recommendations, are accessible through the Internet at http://www.ipm.ucdavis.edu/. Up-to-date pesticide use guidelines, pesticide toxicology information, and other pest management techniques are accessible through this Web site. In addition, you can buy UC IPM pest management publications

SIDEBAR 18

Publications, Treatment Guides, and Pest Management Guidelines from the University of California

The University of California Statewide Integrated Pest Management Program Web site (http://www.ipm.ucdavis.edu) offers information to help you identify and manage turf and landscape pests.

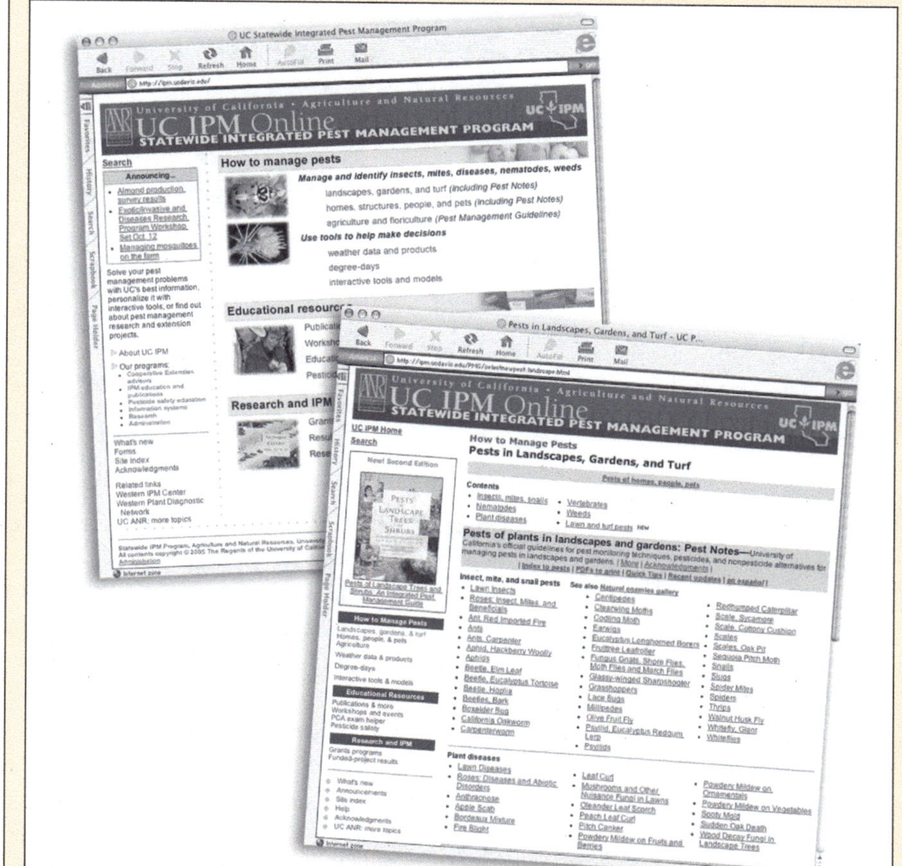

(Figure 8-1) that deal specifically with turf and ornamental plants from ANR Communications Services, 6701 San Pablo Avenue, Oakland, CA 94608. The toll-free phone number for orders is 800-994-8849. A catalog of publications is available upon request, or can be viewed on the Internet at http://anrcatalog.ucdavis.edu. Some UC Cooperative Extension county offices will order these publications for you.

Pest Species

Select a pesticide that is suitable for the pest species you need to control. Make sure the material you choose is effective on the current life stage of the pest. Determine the suitability of the pesticide product by reading the pesticide label. If the target application site is not listed on the label, do not use the material.

Toxicity of the Pesticide

Each pesticide has a signal word that indicates the relative hazard of the pesticide. Factors such as formulation type, persistence in the environment, and amount of pesticide used may modify some hazards. For example,

FIGURE 8-1.

The University of California produces several useful publications for managing pests in turf and landscape.

microencapsulated formulations are generally safer for applicators to handle than are wettable powders.

As a rule—and if you have a choice—select pesticides with the signal word that indicates the lowest level of hazard. A pesticide with the signal word *Caution* is preferable to one with either the *Warning* or *Danger* signal word. If no material with the signal word *Caution* is available, the signal word *Warning* is preferable to a pesticide with the signal word *Danger* (Figure 8-2). A pesticide with the signal word *Caution* will usually be safer to work with, will often be less harmful to people using the area, and will be less disruptive in the environment.

Pesticide Persistence

Depending on the nature of the pest control problem, consider persistence characteristics when you select a pesticide. Persistent pesticides may be more hazardous in areas where people live and work. Persistence also influences the potential hazard should the pesticide move off the target site through leaching or runoff.

The persistence of a pesticide is indicated by its half-life. This is the measure of how long it takes the material to break down to half of the amount of active ingredient originally applied. The longer the half-life, the more persistent the pesticide. Besides the type of pesticide, other factors influence persistence. For instance, the amount of pesticide applied determines how much of the active ingredient remains after a time.

Pesticide formulation types may also affect persistence. Microencapsulated and granular formulations tend to release the active ingredient over a longer period. In this case, only part of the material will begin to break down at the time of application; the rest may begin to break down much later. Pesticides dissolved in oils or petroleum solvents may volatilize more slowly than water-soluble materials, and therefore persist longer. Wettable powders usually have a longer half-life than emulsifiable concentrate formulations.

The pH (see Chapter 3) of the water used for mixing pesticides affects the breakdown speed, and the pH of the

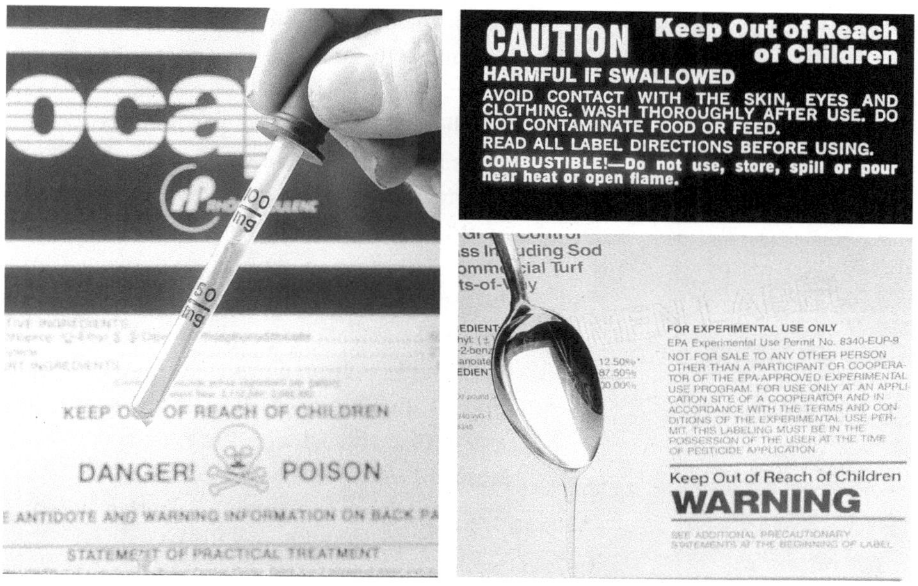

FIGURE 8-2.

Use the signal words on pesticide labels as your guide to the hazards when you select pesticides to use in turf, landscape, and interiorscape settings.

soil or of plant or animal tissues at the application site may have a similar influence. Tissue or soil that is highly alkaline tends to cause more rapid breakdown of some pesticides than neutral or acidic tissue or soil.

The physical nature of the surface being treated influences pesticide persistence. Porous surfaces or soils with high organic matter content adsorb pesticides, reducing the amount of active ingredient available for pest control. For example, areas heavily mulched with organic matter may tie up herbicides used for pre-emergent weed control. In addition, oily surfaces and waxy coatings on leaves and insect body coverings prevent or reduce the uptake of the pesticide. In some cases, these oils or waxes may even combine with the active ingredient, reducing toxicity and persistence.

Soil microorganisms break down many pesticides and influence the persistence of pesticides in the soil environment. These organisms include bacteria, fungi, protozoans, and algae.

Water-soluble pesticides that percolate deeper into the soil break down more slowly than do those remaining near the surface. The main reason for this is that fewer microorganisms are present at greater soil depths. Many pesticides break down more quickly in soils that have a diverse population of microorganisms. A high level of organic matter in the soil often slows pesticide breakdown because of the pesticide's tendency to bind to the organic matter. This makes the pesticide unavailable to the microorganisms. Repeated use of the same pesticide in these highly organic soils can increase the breakdown rate.

Weather also affects persistence. For instance, wind and rain remove or dilute pesticides from target surfaces, lessening their effectiveness. High temperatures and humidity cause chemical changes in some compounds, accelerating their breakdown. Sunlight produces photochemical reactions that decompose many pesticides. Cooler soil temperatures usually increase pesticide persistence because the activity of soil microorganisms decreases.

Cost and Effectiveness of Pesticide Materials

The cost of a pesticide product is an important factor, but you should never base your selection on cost alone.

Check the labels of the products you are comparing to see what rates of the same active ingredient are required, and then convert the cost per pound of active ingredient into cost per unit of area treated. You may discover that you would have to use more of the cheaper product in order to apply enough of the same active ingredient to achieve a level of control comparable to what you would get from a more expensive product. In addition, you must balance your projected cost with the expected degree of effectiveness. A pesticide that costs 30% more but gives 60% better control is often the better bargain.

The effectiveness and quality of a pesticide application can be influenced by weather. Rainfall soon after an application can wash off or dilute sprays, making an application less effective. Windy conditions or indoor air movement during an application will likely produce drift, so coverage of plant surfaces may not be adequate. Some pesticides are more effective at controlling target pests when ambient temperatures are within a certain range. Also, high temperatures or dry and windy conditions may cause some pesticide products to injure the plants.

Effects on Beneficial Insects and Natural Enemies

Always try to conserve beneficial insects and natural enemies. If you are using an integrated pest management program, consider how the selected pesticide will work within the goals of the program. This sometimes means compromising for less immediate control of the pest in order to achieve greater long-term control.

SELECTIVE PESTICIDES

A pesticide's *selectivity* describes the range of organisms that it controls. A *broad-spectrum* pesticide controls a large number of pests and is likely to be harmful to nontarget species whereas a *selective* material controls a smaller number of more closely related organisms, often leaving beneficials and nontarget organisms unharmed.

Some organisms have metabolic ways of altering or detoxifying certain types of pesticides and thereby making these products ineffective against them. In addition, many organisms' behavior is such that it allows them to avoid pesticide exposure.

THE SELECTIVE USE OF PESTICIDES

The way you mix and apply pesticides can improve selectivity (Figure 8-3). Factors that influence the selectivity of many broad-spectrum pesticides include

- application timing
- application technique
- amount of pesticide applied
- type of formulation
- adjuvants used with the pesticide

Application Timing

Proper application timing is important for effective control of target pests as well as for effective protection of natural enemies and beneficial insects. Because some pesticide products are more effective at controlling certain life stages of a given pest, you need to time any application of these materials to coincide with the most susceptible stage.

Whenever possible, avoid any injury to nontarget organisms by choosing an application date or time when they are not present in the treatment area. If you apply insecticides and acaricides while perennial woody plants are dormant, you usually can avoid damage to nontarget and beneficial organisms and still achieve effective control of some plant-feeding mites, aphids, scales, and other insects that overwinter on those plants.

FIGURE 8-3.

The way you apply a pesticide influences its selectivity to the target pest. Proper application methods help to protect natural enemies and make pesticides safer to use.

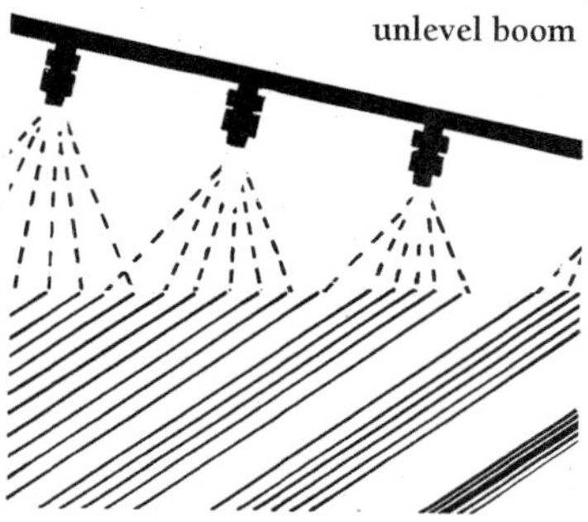

FIGURE 8-4.

When a spray boom with multiple nozzles is tilted, the effect is an uneven spray pattern.

FIGURE 8-5.

Spot treatments increase the selectivity of a pesticide and reduce the amount of pesticide used.

Application Technique

By choosing the correct pesticide application technique, you may be able to reduce drift and lessen hazards to humans and the environment. Application techniques include a number of factors, such as the way you operate application equipment and your use of spot treatments whenever possible.

Equipment Operation. Learn the most effective way to operate your pesticide application equipment. In turf applications, for example, the ground speed must remain constant in order to ensure even pesticide coverage. Nozzles must be properly aligned and must not be clogged, so check and adjust them frequently. To prevent unevenness in the spray pattern, shut off all nozzles during turns. Any rough or uneven ground causes wide booms on equipment to bounce, and this produces an irregular spray pattern. To prevent this, make sure the booms are well supported and parallel to the ground during application. A tilted boom also produces an uneven spray pattern (Figure 8-4). In addition, proper nozzle height is critical for an even spray pattern, so adjust the boom height according to the nozzle manufacturer's recommendation for the type of nozzles being used.

When using a hand-held spray wand, develop a technique that provides a uniform application to all parts of the surfaces being treated. Make sure the spray reaches the upper foliage or branches of trees or shrubs. Direct the spray to all sides of the plants, but avoid excessive runoff. Take care that any upward spraying does not cause the spray to drip down your arm or onto your head.

Walk at a regular pace and avoid uneven steps when operating a backpack sprayer. Hold the nozzle steady and keep it at a uniform distance from the target surface. Maintain even pressure in the spay tank.

Spot Treatments. You can increase selectivity and reduce the amount of pesticide you use by applying the pesticide as a spot treatment rather than as a broadcast application to an entire area (Figure 8-5). For example, some weeds may grow in clumps scattered throughout an area. Spot-treatment involves treating just the weed clumps or patches rather than the whole area. Insects and mites sometimes congregate in a few smaller areas before they disperse over a larger area, especially if the infestation is just beginning. Control these pests by treating before dispersal, and only treating the infested plants. In addition, some pests may occur only on certain plant species. In that case, spot-treat only the infested plants.

Amount of Pesticide Applied

The amount of active ingredient of pesticide applied may have an effect on the selectivity of insecticides and acaricides if your are controlling insect or mite pests. This is because applying lower rates of active ingredient sometimes lessens the impact on natural enemies and other nontarget organisms while reducing pest population numbers. Natural enemies will then be available to control pests that were not controlled by the pesticide.

Type of Formulation

The type of formulation of the pesticide product may influence its selectivity. Granular formulations, for example, do not stick to dry foliage, and this increases their selectivity for soil or aquatic pests. Another type of formulation-related selectivity involves the application of a granular systemic insecticide to the soil. Plant roots take up the insecticide and translocate it to other parts of the plant. Insects that feed on leaves or other plant tissues die, but most natural enemies and beneficial insects do not contact the pesticide and so are left unharmed.

Adjuvants Used with the Pesticide

You can use some adjuvants to improve the selectivity as well as the effectiveness of certain pesticides. Use stickers, spreaders, and drift-control agents to keep spray mixtures on target. Use surfactants to enhance the uptake of pesticides by target pests and attractants to make pesticides attractive to target organisms.

CHAPTER 8: REVIEW QUESTIONS

1. Which of the following would not be a selective way of using a pesticide?

☐ a. making a broadcast application
☐ b. using spot treatments
☐ c. making band treatments
☐ d. using a systemic formulation to protect natural enemies

2. When choosing a pesticide for a specific pest, an important consideration is that it should

☐ a. control the current life stage of the pest
☐ b. repel the pest
☐ c. provide control for more than one year
☐ d. also control natural enemies

3. Pesticides with a high degree of persistence in the turf, landscape, or interiorscape setting are generally

☐ a. recommended over non-persistent pesticides
☐ b. more costly than non-persistent pesticides
☐ c. more hazardous to the public
☐ d. more difficult to apply

4. A factor that may influence pesticide persistence is the

☐ a. type of application equipment
☐ b. age of the plants in the treated area
☐ c. prior pest history
☐ d. formulation type

5. Whenever possible, avoid injuring nontarget organisms while applying pesticides by

☐ a. spraying during the warmest part of the day
☐ b. using strip sprays
☐ c. wearing protective equipment
☐ d. making applications when nontarget organisms are less likely to be present in the area

6. Spot treatments are usually most effective when pest infestations

☐ a. are just starting
☐ b. have become well established
☐ c. do not respond to other application methods
☐ d. have completed their reproductive cycle

7. Which type of an adjuvant would you use to enhance the uptake of the pesticide by the target pest?

☐ a. drift control agent
☐ b. sticker/spreader
☐ c. emulsifiable concentrate
☐ d. surfactant

8. A porous soil, high in organic matter, will likely

☐ a. maximize the amount of applied active ingredient that is available for pest control
☐ b. decrease the amount of applied active ingredient that is available for pest control
☐ c. improve the ability of the pesticide to control the pest
☐ d. make the pesticide more effective, but for a shorter period of time

9 Application Equipment and Calibration

IN THIS CHAPTER we review the different types of pesticide application devices used in landscape, on turf, and in interiorscape settings. We also provide information on the proper calibration of application equipment. For a more in-depth discussion of pesticide application equipment and calibration techniques, see *The Safe and Effective Use of Pesticides, Second Edition.*

APPLICATION EQUIPMENT

Many types of pesticide application equipment are commonly used in landscape, turf, and interiorscape settings. Applicators may use anything from a simple device such as a spray bottle to complex injection units and powered machines.

Applying Liquid Pesticides

Liquid pesticide application equipment is either hand-operated or powered by a mechanical source such as an electric motor or gasoline engine. In general, spraying equipment consists of several components, which include

- a container or tank for holding the diluted pesticide mixture
- a system of pipes or hoses for transferring the pesticide solution from the tank to the nozzles
- a pump or other device for creating pressure to move the liquid
- one or more nozzles for breaking the spray up into small droplets and directing it toward the target
- (on some equipment) pressure regulators, filter screens, control valves, agitators, booms, hoses, and fittings to improve pesticide handling, mixing, and application (Figure 9-1)

Hand-Operated Application Equipment

Hand-operated liquid application equipment is comparatively inexpensive. It is also easy to use. This type of applicator (in most cases, a sprayer) is usually easy to repair and maintain because it has only a few moving parts. You can use these devices to apply pesticides to a small area or to specific plants. They are also handy for locations that are difficult to reach with larger equipment. Hand-operated equipment is lightweight, so a person can easily carry one of these devices. Most are low-pressure sprayers with small tanks. The tanks of some hand-operated sprayers do not have agitators, so they must be shaken from time to time when you use wettable powder, flowable, or emulsifiable concentrate formulations. Table 9-1 is a guide to selecting a hand-operated device that is most suitable for various application needs. Some of the common hand-operated application devices used for controlling pests of turf, landscape, and interiorscape plants are described in the paragraphs that follow. NOTE: When applying insecticides or fungicides to plant foliage, do not use a sprayer that has previously been used to apply herbicides. Herbicide residues in the sprayer may damage the treated plants.

Aerosol Cans. An aerosol can expels a fine spray of ready-to-use (RTU) pesticide formulation through a nozzle

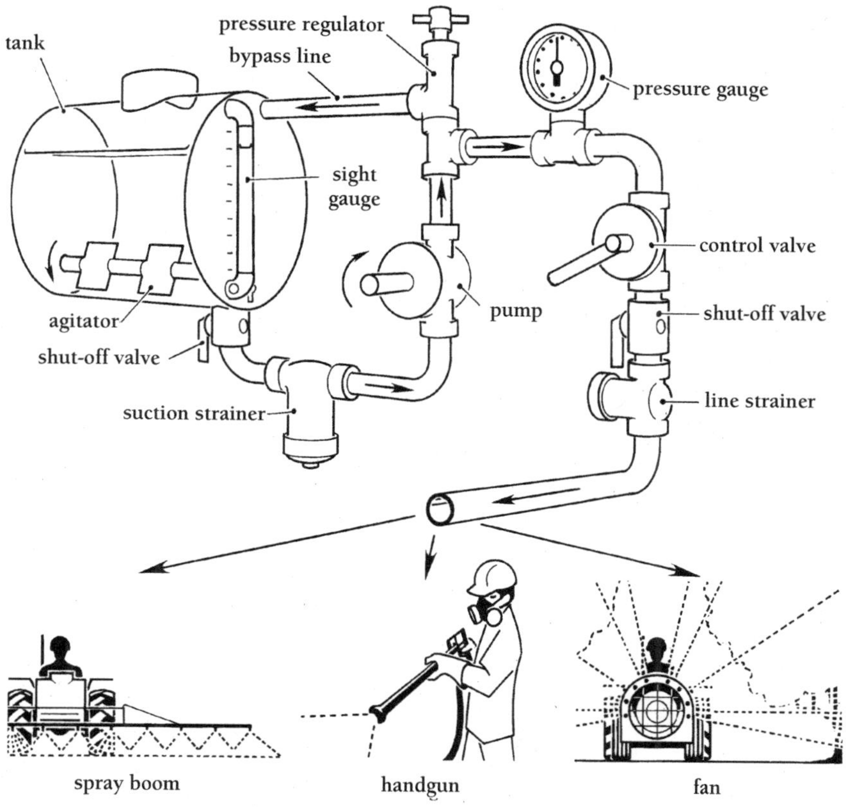

FIGURE 9-1.

Liquid application equipment usually includes a tank for mixing and holding pesticides (often equipped with an agitator), a pump for creating hydraulic pressure, and in some cases a pressure regulator, pressure gauge, control valve, and several types of strainers. Spray is forced through nozzles on a spray boom or handgun, or may be dispersed by a fan.

at the top of the can. The propellant is an inert, compressed gas. Pesticides packaged this way are popular for small applications because of their convenience—no mixing is necessary.

Hose-End Sprayers (Figure 9-2). People commonly use hose-end sprayers to apply pesticides to lawns, flowers, small trees, and shrubs in small to medium-sized outdoor areas. The sprayer combines a pesticide mixture with water from a garden hose and forces the mixture through a high-volume nozzle. A 1- or 2-quart plastic or glass container holds the concentrated or partially diluted pesticide. One filling of the container ultimately produces about 20 gallons of applied diluted spray. On some models, an adjustable nozzle lets you control droplet size and aim the spray in different directions. This type of sprayer also has a valve to start and stop the flow of pesticide in the stream of water. Some have another valve to regulate and shut off the water flow from the garden hose to the pesticide reservoir. Hose-end sprayers can be used to apply insecticides, fungicides, and herbicides. Make sure that the hose or faucet is equipped with an antisiphon valve to prevent backflow of pesticide into the water supply.

Trigger Pump Sprayers (Figure 9-3). The popular trigger pump sprayer is a simple liquid applicator. The trigger-operated pump forces the pesticide mixture through a nozzle, producing a fine spray. Some styles have an adjustable nozzle that allows you to control droplet size and spray pattern. You put a diluted pesticide mixture into

TABLE 9-1.

Selection Guide for Nonpowered and Hand-Operated Application Equipment for Liquid Pesticides.

TYPE	USES	FORMULATIONS	COMMENTS
Aerosol can	Insect control on houseplants or patio plants, pets, small areas, cracks and crevices, and confined spaces.	Liquids must dissolve in solvent; some dusts are available.	Very convenient. High cost per unit of active ingredient.
Backpack sprayers	Many commercial and homeowner applications. Can develop high pressures. Used for insect, weed, and pathogen control. Often used indoors for household pest control.	All formulations. Wettable powders and emulsifiable concentrates require frequent shaking.	Durable and easy to use. Requires periodic maintenance.
Compressed air sprayers	Same uses as backpack sprayers.	All formulations. Wettable powders and emulsifiable concentrates require frequent shaking.	Good overall sprayer for many types of applications. Needs thorough cleaning and regular servicing to keep sprayer in good working condition and prevent corrosion of parts.
Hose-end sprayer	Home garden and small landscaped areas. Used for insect, weed, and pathogen control.	All formulations. Wettable powders and emulsifiable concentrates require frequent shaking.	Convenient and low-cost way of applying pesticides to small outdoor areas. Cannot spray straight up.
Trigger pump sprayer	Indoor plants, pets, and small home yard areas. Used for insect and pathogen control.	Liquid-soluble formulations best.	Low cost and easy to use.
Trombone sprayer or slide sprayer	Used for treating shrubs and medium-sized trees up to about 25 feet tall.	All formulations. Wettable powders and emulsifiable concentrates require frequent mixing.	Applies uniform concentration of active ingredient if properly mixed. Relatively easy to clean and maintain. Moderately priced. Effort required to maintain pressure by continually operating a push-pull slide handle. Requires practice to obtain uniform spray coverage.
Wick applicators	Used for applying contact herbicides to emerged weeds. Agricultural and landscape uses.	Only water-soluble herbicides.	Simple and easy to use. Clean frequently.

the plastic container, which ranges in capacity from 1 pint to 1 gallon. Use this type of applicator to apply insecticides or fungicides to a small area or to specific plants, such as in interiorscape settings. Manufacturers package some ready-to-use herbicides and other types of pesticides in trigger pump sprayers. Before adding a pesticide mixture, check the operation of the trigger sprayer with water to make sure that it works properly and that the nozzle is not clogged.

Compressed-Air (Pump) Sprayers (Figure 9-4). Compressed air sprayers hold a diluted pesticide mixture in a small, airtight tank. The tanks have a capacity of 5 gallons or less. Common tank sizes include 1 quart and ½, 1, 2, and 3 gallons. The sprayer has a hand pump that compresses the air inside the tank. The compressed air forces the liquid pesticide solution through a hose and nozzle when you open a valve (see the section below on spray nozzles). Some types use compressed carbon dioxide cartridges as the propellant, eliminating the need for hand pumping. These cartridges provide more even pressure than the hand pump.

Use compressed air sprayers to apply herbicides, insecticides, and fungicides to small outdoor areas and to apply insecticides or fungicides to plants

FIGURE 9-2.

Hose-end sprayers are sometimes used to apply pesticides to lawns, shrubs, and small trees. They apply high volumes of diluted pesticide.

FIGURE 9-3.

A trigger-pump sprayer can be used to apply small quantities of diluted pesticide in interiorscapes and for some types of herbicide applications.

indoors. Select a unit that has an adjustable nozzle so you can control droplet size and spray pattern.

Backpack Sprayers (Figure 9-5). Backpack sprayers have a hand-operated hydraulic pump that forces liquid pesticide through one or more nozzles attached to a hand-held wand (see the section below on spray nozzles). You operate the pump by moving a lever up and down and you control the spray with a squeeze valve on the spray wand. Some pumps can create a pressure of more than 100 pounds per

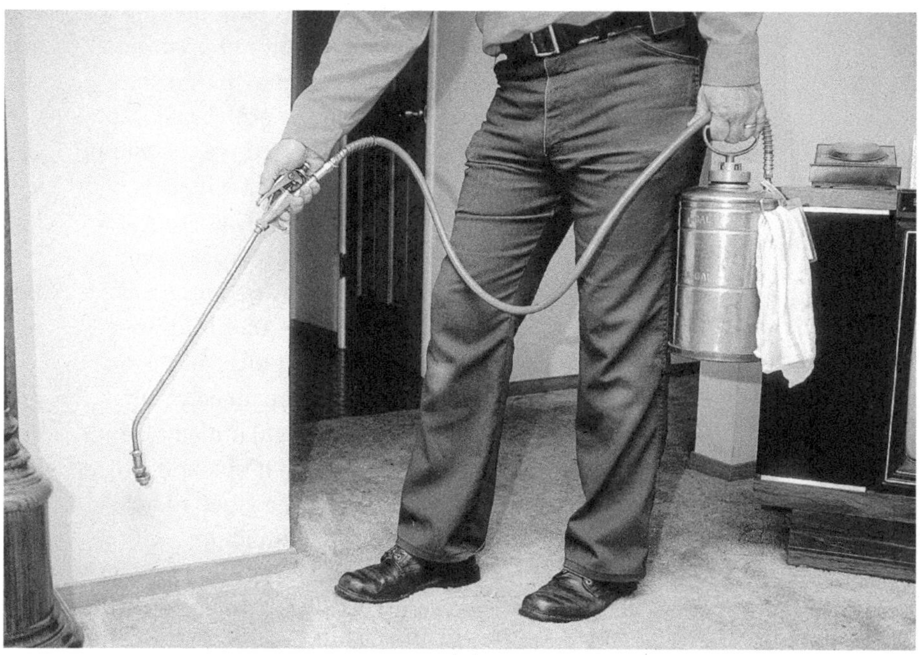

FIGURE 9-4.

Compressed-air sprayers usually hold from ½ to 5 gallons of spray mixture. Air inside the tank is compressed with a self-contained pump or a carbon dioxide cartridge. Diluted pesticide under pressure is forced through a hose attached to an adjustable nozzle at the end of a hand-held wand.

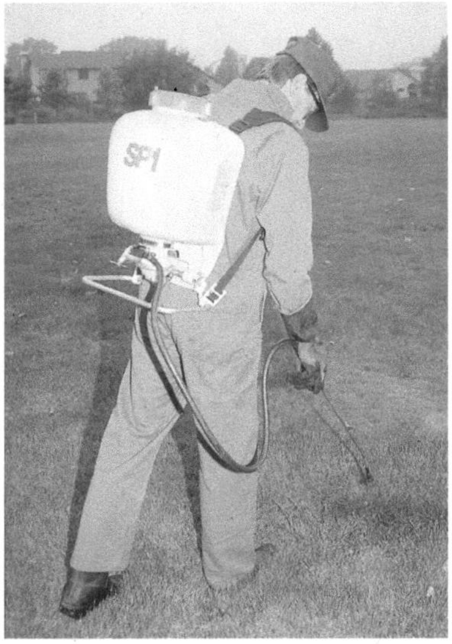

FIGURE 9-5.

This type of hand-operated backpack sprayer usually requires continuous pumping action to maintain pressure for spraying. These units hold from 3 to 5 gallons of spray mixture. Most have adjustable nozzles.

FIGURE 9-6.

A wick applicator is used for application of contact herbicides. It can be used in areas where weeds are taller than the landscape plant. Wick applicators reduce problems of drift and waste of herbicides.

square inch (psi). The plastic tanks have a capacity of around 5 gallons. These units are useful for spraying small to medium-sized areas. You can use them to apply herbicides, insecticides, and fungicides.

Rope Wick or Pad Applicators (Figure 9-6). Use rope wick applicators or pad applicators to apply systemic herbicides for weed control. The basic design consists of a pad made of cloth or carpeting or a rope wick. The unit saturates the pad or wick with herbicide mixture. You wipe the pad or wick onto the leaves of weeds. A simple hand-held model holds liquid herbicide in its hollow handle. Liquid feeds to the pad or wick through small holes. This device is effective for applying the herbicides to weed leaves. It does not have the drift problems associated with spray applications.

Powered Application Equipment

Turfgrass areas and other larger-scale pesticide application situations require powered equipment that is capable of covering a large outdoor area more quickly than hand-held equipment can. Some spray units have their own dedicated electric motor or gasoline engine to drive the spray pump and others depend on the power take-off from a small tractor to drive the pump. Manufacturers equip some models with hydraulic or mechanical agitators and pressure regulators. Units also come with a variety of spray booms, hand spray guns, and other nozzle arrangements. Powered equipment is more complex than hand-operated equipment and requires regular maintenance and servicing to keep it operating properly. Table 9-2 is a guide to help you select powered liquid-application equipment.

Powered Backpack Sprayers. The smallest powered sprayer consists of a backpack unit with a compact gasoline engine. The engine drives a pump that forces pesticide mixtures through one or more nozzles (see the section below on spray nozzles). A backpack sprayer works best for low-volume applications. This is because it has a small tank and cannot generate a high-pressure spray. Use this type of sprayer to apply herbicides, insecticides, or fungicides. If you are applying insecticides or fungicides to plant foliage, do not use a powered backpack sprayer that has previously been used for herbicide applications unless you first clean and thoroughly decontaminate the tank and hoses.

Low-Pressure Sprayers (Figure 9-7). A low-pressure pesticide sprayer is a useful tool in many different turf, landscape, and median strip settings. This equipment operates in the pressure range of 20 to 40 psi with a battery-powered electric motor or small gasoline engine to operate the pump. Most low-pressure sprayers have a tank with a capacity of 30 to 100 gallons or more. For turf applications, this type of unit may be towed behind a small all-terrain vehicle (ATV), a small tractor, or some other type of vehicle. You can use some types to inject pesticides into the soil. Multiple nozzles (see the section below on spray nozzles) can be arranged along a boom to spray a wide swath.

High-Pressure Hydraulic Sprayers (Figure 9-8). High-pressure sprayers work well on turf areas, trees, and large shrubs. This equipment forces a high volume of diluted pesticide either through a hand spray gun or through nozzles mounted on a boom .

Pressures of 100 psi and higher are common in high-pressure sprayers. Most sprayers of this type have a large-capacity tank. A high-pressure hydraulic sprayer mounts onto a trailer, truck, or tractor and may include a gasoline engine that powers the pump. Many models have a pressure regulator and bypass mechanism that allow them to work as a low-pressure sprayer as well.

Use high-pressure sprayers to apply insecticides and fungicides. Because of

TABLE 9-2.

Selection Guide for Powered Liquid Pesticide Application Equipment.

TYPE	USES	FORMULATIONS	COMMENTS
Controlled droplet applicators	Used for application of contact herbicides and some insecticides. Some applicators are hand-held while others are mounted on a spray boom.	Usually water-soluble formulations.	Plastic parts may break if handled carelessly.
High-pressure hydraulic sprayer	Landscape, right-of-way, and agricultural applications. Use on dense foliage and large trees or shrubs.	All. Equipment may include an agitator.	Useful for larger areas. Important to clean and service equipment frequently. Sprayer requires its own motor or external power source. Abrasive pesticides may cause rapid wear of pumps and nozzles.
Low-pressure sprayer	Very common type of sprayer used in commercial applications for weed, insect, and pathogen control. Used with spray booms or hand-held equipment.	All. Equipment may include an agitator.	Useful for larger areas. Frequent cleaning and servicing required. Powered by own motor or external power source.
Powered backpack sprayer	Aquatic, landscape, right-of-way, forest, and agricultural applications.	All. Some may require agitation.	Applies a uniform concentration of active ingredient. May be too heavy to carry for long periods of use. Requires frequent maintenance.

the high pressure of the spray, drift may be a problem.

Electrostatic Sprayers. An electrostatic sprayer applies pesticide in the form of small, electrically charged droplets. The droplets average about 50 microns in diameter and receive a negative electrostatic charge as they leave the sprayer. Because plant material has a positive electrostatic charge, plant surfaces attract the spray droplets. The negatively charged spray droplets repel each other, so they do not clump together to form larger droplets. These sprayers appear to be more effective if the spray droplets only travel a short distance between the sprayer to the target. As the distance increases, the effect of the electrostatic charge diminishes.

FIGURE 9-7.

A common sprayer used for turf and some landscape applications of pesticides is the low-pressure applicator. It apply spray through a series of nozzles attached to a boom.

FIGURE 9-8.

High-pressure sprayers work well for spraying turf areas, trees, and large shrubs. This equipment forces high volumes of diluted pesticide either through a hand spray gun or through nozzles mounted on a boom.

Hand Spray Guns

The hand spray gun is a component of certain types of spraying equipment. It usually consists of a handle, a valve, and a nozzle. The valve is part of a trigger mechanism or connects to a knob at the end of the handle. Most hand spray gun nozzles are interchangeable to allow you to customize the spray for different types of applications. The valves on some models allow you to adjust the pressure and spray pattern.

Use hand spray guns to apply a high-pressure stream or spray of insecticide or fungicide to trees and shrubs (Figure 9-9).

Spray Nozzles

The spray nozzles control the application rate, droplet size, and spray pattern of a pesticide application. They also contribute to the thoroughness and safety of the application (Figure 9-10). Several different types of nozzle are used depending on the type of application and the application equipment. Nozzles are among the most important parts of the sprayer. If you do not carefully select and maintain the nozzles, you can waste all of your efforts for effective pest control. Base your selection of the right nozzle on several criteria, including

- type of material the nozzle is made from
- style of the nozzle
- nozzle orifice size

Spray Nozzle Construction and Wear. Manufacturers make nozzles from several different materials, all of which are subject to wear. The design of the nozzle, the pesticide formulations being applied, and the spray pressure all influence nozzle wear. A flat-spray nozzle with a sharp-edged orifice will wear much more quickly at first than, for example, a flooding tip with a circular orifice. As the spray pattern angle increases, the wear on the nozzle increases. Furthermore, the size of the orifice affects wear: a larger orifice wears more slowly than a smaller one.

Pesticide product formulations affect wear to an extent that varies with the amount of dissolved or suspended solids in the liquid. A true solution (where the pesticide is actually dissolved and not just emulsified or suspended) cause

FIGURE 9-9.

Hand spray guns usually produce high-pressure streams or sprays; they are used for applying insecticides and fungicides to turf, trees, and shrubs.

FIGURE 9-10.

Spray nozzles control the application rate, drop size, uniformity, thoroughness, and safety of the pesticide application.

the least amount of wear, while suspended solids such as wettable powders cause various degrees of wear. Wear caused by suspended solids depends on the particles'

- size
- range of sizes (for non-uniform suspended particles)
- shape
- hardness
- concentration

As a nozzle wears, the volume and pattern of its spray change, and this affects the accuracy of application. Replace nozzles when they fail to deliver the desired amount of spray volume or do not produce the spray pattern described by the nozzle manufacturer. The output from any two nozzles of the same size used together on a boom should not vary one from the other by more than 10%. To ensure uniform wear, make sure that all nozzles on a boom are made from the same material. Manufacturers produce nozzles from the following materials:

- *Brass.* Brass nozzles are moderately inexpensive but wear quickly from abrasion. Brass is an acceptable material if you do not use abrasive sprays or if you replace nozzles frequently.
- *Stainless steel.* Stainless steel nozzles do not corrode and they resist abrasion. Although hardened stainless steel wears exceptionally well, these nozzles are more expensive than most other nozzle types.
- *Aluminum and monel.* Aluminum and monel nozzles resist corrosion but are highly susceptible to abrasion because they are such soft metals. Avoid using aluminum or monel nozzles unless you specifically need resistance to corrosion.
- *Plastic.* Plastic nozzles are the least expensive. The plastic material resists corrosion, but nozzles made entirely of plastic may swell if exposed to organic solvents. Plastics also have a low resistance to abrasion. Use solid plastic nozzles only with water-soluble pesticides. Some plastic nozzles are fitted with a stainless steel orifice insert, making them much more resistant to wear. The inserts also reduce problems associated with swelling.

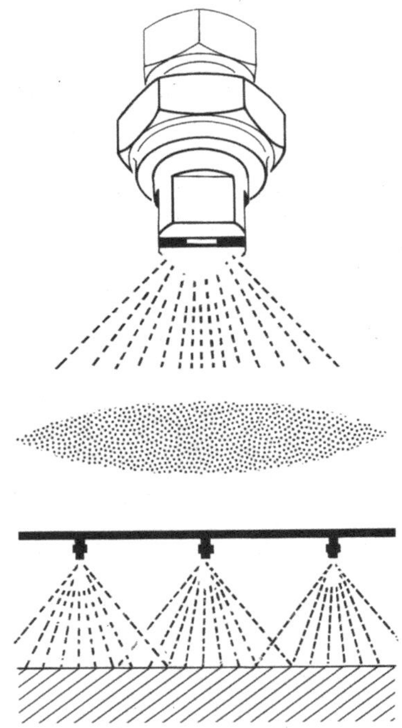

FIGURE 9-11.

Flat-spray nozzles produce a fan-shaped pattern that has more droplets in the center part of the fan than at either edge. This allows for overlap of the spray, eliminating gaps and providing an even pattern when multiple nozzles are used.

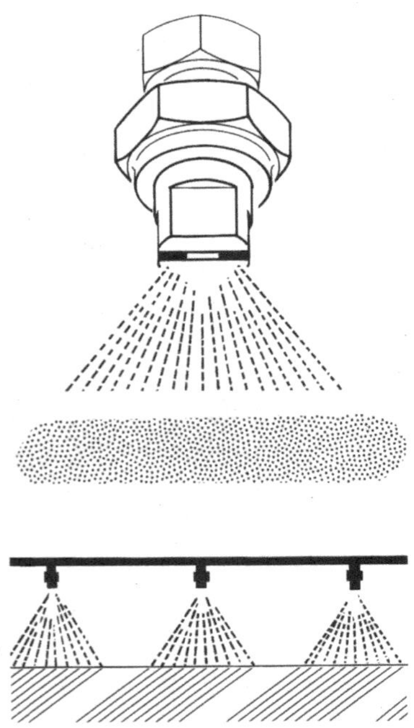

FIGURE 9-12.

Even flat-spray nozzles provide a uniform distribution of spray throughout the fan pattern. The spray from these nozzles does not overlap. These nozzles are used to apply separate bands of pesticide without overlap.

- *Tungsten carbide and ceramic.* Tungsten carbide and ceramic nozzles are highly resistant to abrasion and corrosion. To reduce costs, manufacturers use tungsten carbide or ceramic inserts with brass or plastic nozzle bodies. Use this type of nozzle for high-pressure and abrasive sprays.

Nozzle Types. Different application situations require nozzles designed for specific spray pattern needs. A nozzle that you use to apply herbicides onto a sports field may be unsuitable for applying insecticides or fungicides to plant foliage. Interiorscape applications require nozzles that are suitable for use in confined spaces.

Flat-Spray Nozzles (Figure 9-11). A flat-spray nozzle distributes pesticide in a flat fan shape, with the angle of the fan's outside edges ranging between 50 and 160 degrees. Flat-spray nozzles produce a pattern with more spray droplets in the center of the fan. The pattern tapers off at each end. To achieve even coverage, you space a series of flat-spray nozzles evenly along a boom, allowing the edges of the spray pattern from each nozzle to overlap those of its neighbors. Manufacturers recommend that you rotate all nozzles 10 degrees from the axis of the boom to keep droplets from adjacent nozzles from colliding and still allow for proper overlap of the spray pattern (see Figure 9-35). If you operate the nozzles at the correct height, the spray patterns will combine into an even swath with nearly uniform deposits between nozzles. Use several flat-spray nozzles spaced along a boom for even application of herbicides, fungicides, and insecticides on turf or other large areas.

Low-pressure flat-spray nozzles provide an acceptable spray pattern at pressures as low as 10 psi. Use these nozzles for the same types of applications as you would cover with conventional flat spray nozzles. The low-pressure nozzles have the advantage, though, of having fewer problems with drift because they produce larger droplets at a lower pressure.

Even Flat-Spray Nozzles (Figure 9-12). Manufacturers produce even flat-spray nozzles in 40-, 80-, and 95-degree fan angles. They are similar to flat nozzles, but there is no tapering of spray volume at the pattern ends. Use these on spray wands powered by compressed air or on backpack sprayers when applying one or more separate bands that will not overlap.

Cone Nozzles (Figure 9-13). Use cone nozzles to apply insecticides and

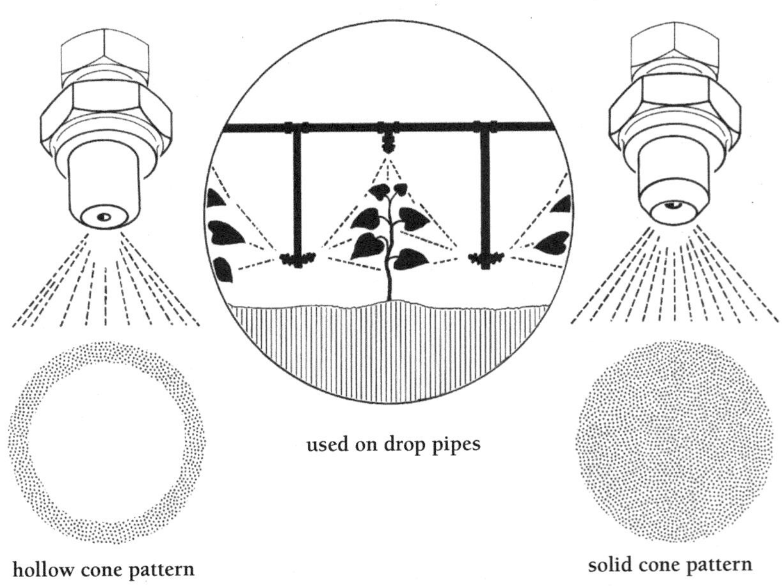

FIGURE 9-13.

Cone nozzles are used for applying insecticides and fungicides to foliage, especially when larger volumes are required to ensure complete coverage.

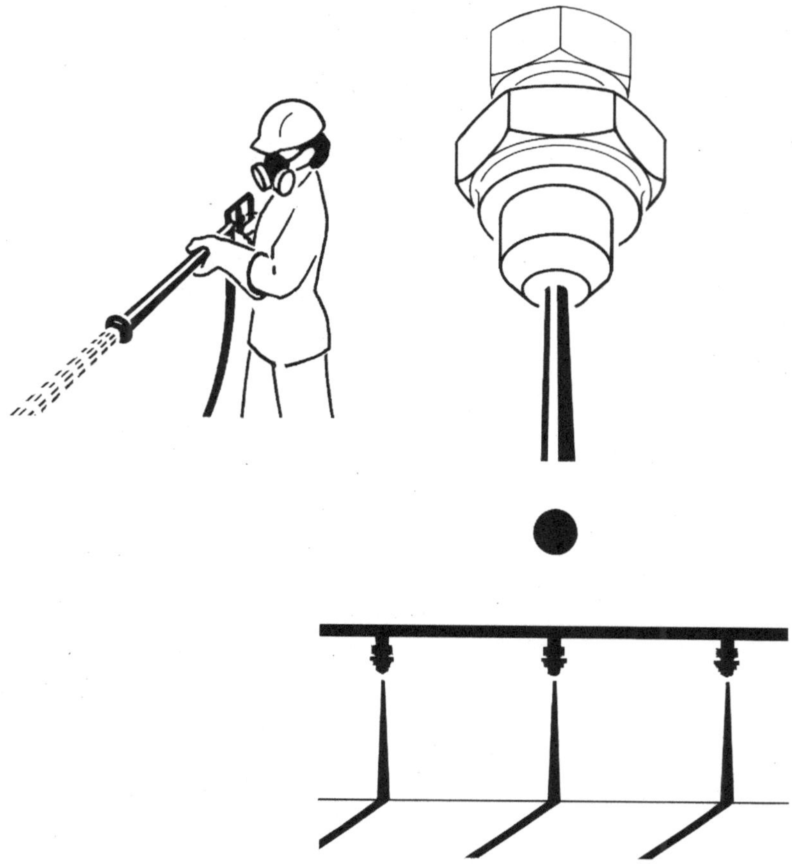

FIGURE 9-14.

Solid-stream nozzles are used in high-pressure handguns and in low-pressure crack and crevice units. They are also used on boom sprayers to apply fluids in bands.

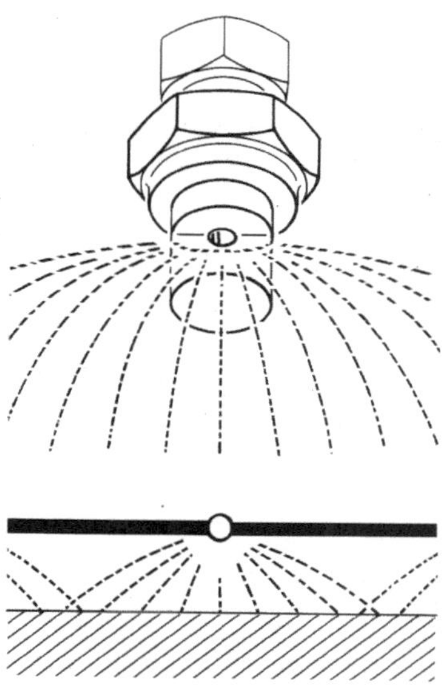

FIGURE 9-15.

Flood nozzles are used to apply large volumes of liquid under low pressure. They are occasionally used for pesticide application, but are more commonly used for liquid fertilizers.

fungicides to dense foliage using a compressed air or backpack sprayer. A cone nozzle produces spray in either a hollow-cone or a solid-cone pattern, with the angle of the outside edges of the cone varying between 20 and 110 degrees. Use hollow-cone spray nozzles for most applications.

Solid-Stream Nozzles (Figure 9-14). A solid-stream nozzle produces a single, solid stream of pesticide. Use this type of nozzle on hand spray guns used for spraying trees and large shrubs. The orifice size determines the output volume.

Flood Nozzles (Figure 9-15). A flood nozzle produces a relatively wide fan angle, up to 160 degrees. Use flood nozzles to apply large volumes of liquid at a low pressure. This nozzle type works well for liquid fertilizers when you need to apply a large volume of liquid. Pest managers rarely use flood nozzles for pesticide application, though, because pesticides usually do not require application of such large volumes of liquid.

Nozzle Tip Numbers. Most manufacturers use a method of numerical coding to identify nozzle tips and engrave the appropriate code number onto the face of the nozzle. For example, "8004" is one manufacturer's code for a common flat spray nozzle. The first two digits indicate that the nozzle produces an 80-degree fan spray. The last two numbers indicate the volume of spray output (0.4 gallons per minute) at 40 psi (the manufacturer's recommended spray pressure). A nozzle with the code number 65155 is a 65-degree fan spray producing a volume of 1.55 gallons per minute at 40 psi. Check individual manufacturers' catalogs to determine the rated operating pressure for the nozzles you want to use. Some nozzle styles operate at higher or lower pressures.

When selecting nozzles, learn about the nozzle types and sizes, their proper application, and their optimum pressure range recommendations. Manufacturers have developed charts to help you select the right nozzle type and size to achieve your desired spray volume. Table 9-3 is a guide for selecting the appropriate nozzle style for an application.

Applying Granules

Granules are the particles of a dry formulation used for plant pest control. You usually use water to incorporate granules into the soil after application. Table 9-4 is a guide for selecting granule application equipment.

Granule applicators may be hand-operated, mechanically driven, or engine powered, depending on the needs of the application site. Because granules vary in size and shape, the application equipment must be able to accommodate these differences. In a pellet formulation, the granules are of identical size and shape. Uniform size and shape makes it easier to calibrate the application equipment and make an even application.

Hand-Operated Granule Applicators. Most hand-operated granule application devices strap to the operator's chest. A common name for this type of applicator is *belly grinder.* Granules pass through an adjustable opening at the bottom of a metal or plastic hopper. They drop onto a spinning plate operated by a hand crank. The operator walks at a steady pace while turning the crank to achieve an even distribution of granules.

TABLE 9-3.

Guide to Selecting Different Nozzle Styles.

STYLE OF NOZZLE	SUGGESTED USES	RECOMMENDED PRESSURE	TYPE OF SPRAY PATTERN
Broadcast nozzles	Weed and brush control in pastures and turf. Nozzles are clustered and are not used without a boom.	10 to 30 psi.	Wide, fan-shaped pattern with droplets that range from fine to coarse.
Cone nozzles	Insecticides and fungicides applied	40 to 120 psi.	Hollow or solid cone pattern. Fine spray droplets, good penetration.
Even flat-spray nozzles	Preemergence and post-emergence herbicides, insecticides, and fungicides. Use on a boom. Do not overlap the spray pattern.	20 to 40 psi. Keep pressure low when used for weed control.	Fanlike pattern with even distribution of spray across the width of the fan.
Flat-spray nozzles	Preemergence and post-emergence herbicides, insecticides, and fungicides. Used on a boom.	20 to 60 psi. Keep pressure as low as possible when spraying weeds.	Fanlike pattern with fewer droplets at sides than in center of fan pattern. Suitable for overlapping with other nozzles to produce a wide spray swath.
Flood nozzles	Herbicides and fertilizers. High volume and low pressure to reduce drift. Used on booms.	5 to 20 psi.	Wide, fanlike pattern of coarse droplets.
Off-center flat-spray nozzles	Used on ends of a spray boom to extend the reach of the spray pattern.	Same as flat-spray nozzles.	Fan-shaped, with angle to one side.
Solid-stream nozzles	Pesticides of all types. Used on booms or hand guns.	5 to 200 psi.	Low- or high-pressure solid stream. High pressure breaks the spray into fine to medium droplets.

TABLE 9-4.

Selection Guide for Granule Application Equipment.

TYPE OF APPLICATOR	USES	FORMULATIONS	COMMENTS
Hand-operated granule applicator	Landscape, aquatic, and some agricultural areas.	Granules or pellets.	Suitable for small areas. Easy to use. Inexpensive. Can be difficult to apply a uniform, accurate amount of active ingredient.
Mechanically driven granule applicator	Turf and other landscape areas. Also commonly used in agricultural areas.	Granules or pellets.	Requires accurate calibration.
Powered granule applicator	Agricultural areas—usually row crops. Some large turf or landscape applications.	Granules or pellets.	Frequent servicing and cleaning are required.

FIGURE 9-16.

The typical drop-type spreader has an agitator bar mounted on the spreader axle. This bar rotates over a series of openings in the bottom of the spreader hopper, and granules exit the hopper through these openings. The size of the openings can be adjusted to obtain different rates of application.

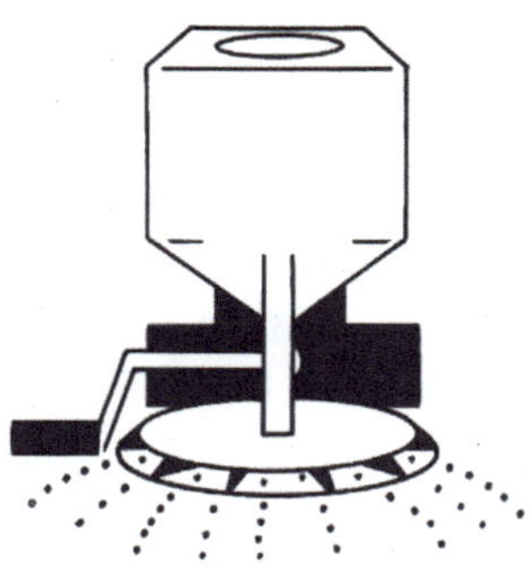

FIGURE 9-17.

Rotary granule applicators vary in size to accommodate a range of application needs. One or more openings at the bottom of the hopper let granules drop onto a spinning disc. These applicators are ground-wheel-driven, meaning one of the unit's turning wheels is the source of power for the spinning disc as you push or tow the unit.

Drop Spreaders. The typical drop-type spreader (Figure 9-16) has an agitator bar mounted on the spreader axle. This bar rotates above a series of openings in the bottom of the spreader hopper, and granules exit the hopper through these openings. The openings' size can be adjusted to vary the rate of application. Because all of the granules fall within the spreader's wheelbase, you have to overlap passes slightly in order to ensure uniform coverage. Common problems associated with drop spreaders include

- skips
- insufficient or excessive overlap
- variable spreading when turning corners
- clogged openings in the bottom of the hopper if turfgrass is moist during the application or if granules clump together

Drop spreaders are generally more precise than rotary spreaders and deliver a better pattern. In addition, because the granules drop straight down and only a short distance, there is less chemical drift and pattern distortion. To avoid problems when turning corners, always turn off the spreader while you are making a turn. This means that you should first treat the strips at both ends of the area (the turning zone). That way you can start out in the treated end strip and then turn the spreader on as you cross into the untreated area.

Rotary Granule Applicators (Figure 9-17). Several types of rotary granule applicator are available. These vary in size to accommodate a range of application needs. One or more openings at the bottom of the hopper lets granules drop onto a spinning disc. This disc scatters the granules over a wide area. These applicators are ground-wheel-driven, meaning that one of the unit's wheels powers the spinning disc as you push or tow the unit. By adjusting the size of the opening that granules pass through, you can control the flow of granules onto the spinning disc and regulate the rate of application. Your travel speed controls the swath width since the spinning disc turns faster as travel speed increases and granules are scattered more widely.

Common problems associated with rotary spreaders include the following:

- A dry product with varying granule sizes and weights will not spread uniformly.
- The amount of overlap required to obtain a uniform pattern differs for each product, since each product has a slightly different particle size and weight.
- The opening at the bottom of the hopper often gets clogged.
- Any turning changes the rotation speed of the plate, causing uneven distribution of granules.

Rotary spreaders are generally less precise than drop spreaders in terms of

FIGURE 9-18.

Granules can also be applied with a backpack granule applicator. A small gasoline engine powers a blower that disperses the granules.

FIGURE 9-19.

Drenching soil at tree base with a systemic insecticide.

uniformity of distribution. You can correct the tendency toward non-uniform distribution of granules by properly overlapping adjacent swaths. Always try to minimize the drift of small particles when using a rotary spreader.

Rotary spreaders throw material 3 to 4 feet in front of the spreader. The operator is another 3 to 4 feet behind the spreader's hopper. This means that when the material finally reaches the turf, it is 6 to 8 feet in front of the operator. It is essential then that the operator run a border swath along the edges of the area being treated so there will be enough room to maneuver the spreader for the next pass.

Because rotary spreaders throw material in a swath 6 to 8 feet wide, granules are likely to be spread onto sidewalks, driveways, patios, and other nontarget surfaces. Should this occur, it is important that you sweep the granules off of these surfaces and back onto the treatment area.

Remember to open and close the spreader at the beginning and end of each pass. Start walking before you open the hopper's application ports and then make sure to close them before you stop walking. If you keep the spreader open while turning around you will cause a misapplication: since the spinning disc is driven by the rotation of one of the spreader's side wheels, it will stop rotating when you turn the spreader around in one direction (toward the drive wheel), and turn faster than normal when you turn the spreader in the other direction (away from the drive wheel).

Powered Granule Applicators (Figure 9-18). A small gasoline engine provides the power to operate a backpack granule applicator. This type of applicator is similar in size and weight to a powered backpack liquid sprayer. The engine operates a blower that propels granules through a flexible hose. The operator aims the hose at the target area while walking slowly. It is difficult to get an even distribution of granules with this type of powered applicator, so it is important that you maintain a uniform walking speed and keep the discharge hose steady in order to keep the application rate relatively uniform throughout the application.

Soil Drenches and Soil Injection

Soil drenches and soil injection are effective ways to apply certain systemic pesticides to small numbers of trees or shrubs. These application methods are especially useful in populated areas because they reduce the risk of exposure to people and eliminate the possibility of drift. You can also use drenches and soil injection in some interiorscape settings. For applications to trees growing along city streets and other woody plantings in highly populated areas, soil injection or soil drenches may provide a safe and practical alternative to foliar sprays (Figure 9-19).

You can make a soil injection or soil drench treatment using either powered or hand-held equipment. You can adapt any equipment that can operate at low pressure for this sort of application. Typically, such treatments require pressures of less than 25 psi. Powered equipment for soil injection consists of a tank sprayer fitted with a hose and an injector probe (Figure 9-20). Hand-held equipment is practical only if you are treating a limited number of trees or shrubs, since a hand-held device can hold only a small volume of diluted pesticide.

Tree Injection and Implantation

Many people know about the age-old practice of hammering galvanized nails into the trunk of a tree in order to provide the tree with zinc. Tree injection is based on a similar principle: a substance is injected into the sapwood of a tree by way of a syringe-like injector. The injector is attached to a

FIGURE 9-20.

Soil injection (as an alternative to injecting or implanting trees) is possible even if the trees are mostly surrounded by pavement, so long as it is permitted on the product label.

feeder tube that leads either to a small tank of pesticide or, more commonly, to pre-measured capsules of the substance being injected.

Tree implantation and injection systems supply nutrients and apply systemic fungicides and insecticides. Using this method, you can use a systemic pesticide to control existing pest problems or to protect a plant against future infestation. Injection or implantation can also be used to place herbicides into the vascular system of an undesirable tree or shrub that you wish to eliminate from the landscape.

These are relatively labor-intensive pest control techniques and are expensive both in terms of materials and of time, and the numbers of pests controlled in this manner are few. The biggest advantage to injection or implantation is that there is little chance that drift or other possible public hazards will result from the pesticides applied. Soil drenches and soil injections, discussed above, have the same advantage, but often have fewer of the disadvantages associated with tree injection or implantation.

Tree Injection

There are two types of tree injection techniques: macroinjection and microinjection. With macroinjection, you place the pesticide mixture into a container on the injection applicator. Small holes bored into the tree trunk provide access to the outer layers of sapwood. You run tubing into the holes and connect the tubes to a manifold on the container that holds the pesticide mixture. The application device forces pesticide into the living tissue (xylem) under the bark where it is absorbed. Injection takes several hours, and you must remain in the area to monitor the process until it is completed.

Microinjection systems also require that holes be drilled into the tree, but the holes used for microinjection are smaller and shallower than those used for macroinjection. The microinjection system consists of a small plastic capsule containing the pesticide attached to a short plastic connector called the "feeding tube." You insert the tube into a predrilled hole at the base of the trunk or in some cases at the root flare or between root buttresses. Each capsule is pressurized. By tapping the capsule with a mallet you break a seal, releasing its contents through the feeding tube and into the plant's xylem. After injection, you remove the capsule and feeding tube.

Microinjection's pesticide delivery system depends on the negative pressure in the xylem that occurs when trees transpire and lose water. Transpiration creates the negative pressure that pulls xylem fluids upward through the tree. Microinjection will not work on a dormant trees, since there is no movement of xylem fluids.

Tree injection is a fairly expensive and time-consuming form of pest control and is only practical for situations where you are protecting valuable trees that have been growing for a long time. Injection is only effective for a few types of pest. One further disadvantage of tree injection is that it creates visible wounds in tree trunks and those wounds can serve as an entry path for new pathogens and some insects, jeopardizing the health, structural stability, and appearance of treated trees (Figure 9-21).

Implantation

Implantation involves the insertion of solid chemicals, usually contained in a water-soluble capsule, into holes that you drill into a tree's outer sapwood. You must make sure that the outer end of the capsule is flush with the outside of the tree's bark. Placed in the outer sapwood, the capsule will be dissolved by the tree's sap, which will absorb the chemical it contains. The sap stream carries the chemical upward into the tree.

FIGURE 9-21.

One disadvantage of tree injection is that it creates visible wounds in trunks and can provide an entry path for pathogens and some insects. This can jeopardize the health, structural stability, and appearance of treated trees.

This method only allows you to apply small amounts of chemical, and it only works with pesticides that will dissolve in the plant sap. One advantage to this technique is that the chemical usually is dissolved and distributed slowly over time, and provides long-term pest control.

Like tree injection, implantation can cause visible wounds in treated trees and so may provide an entry route for invading pathogenic organisms.

FIGURE 9-22.

Poisoned baits are usually placed in tamper-resistant bait stations or protected in some other way to prevent children or nontarget animals from being exposed to the toxic material. When using bait stations, be sure they are properly labeled and secured.

Bait Application Equipment

Pesticide bait formulations require special application methods in order to keep the bait fresh and attract the target pest. A major problem associated with baits is that they can attract and expose nontarget organisms to toxic pesticides. Bait stations or specialized bait application devices can help prevent such exposure.

Bait Stations

A bait station holds a supply of poisoned grain or other food that attracts target pests. Use tamper-resistant styles that block access to the bait by children, pets, and nontarget animals (Figure 9-22). Manufacturers make bait stations for control of squirrels, rodents, cockroaches, ants, snails, and other pests.

Bait Application Devices for Gophers and Moles

There are commercially available application units for applying poisoned baits to control gophers and moles in turf and landscape areas. Hand-operated models inject a small quantity of poisoned bait directly into the underground burrow made by the gopher or mole. For large areas of turf, a tractor-mounted mechanical bait applicator will form artificial burrows that intersect with natural gopher burrows (Figure 9-23). This equipment inserts bait at regular intervals inside the artificial burrow. Gophers exploring the artificial burrows feed on the bait.

Trapping for Rats, Mice, and Other Small Animals

Whenever possible, use mechanical or sticky traps to control rodents and

FIGURE 9-23.

This mechanical bait application device is used to apply poisoned bait for control of gophers; it forms an artificial burrow that intersects with the natural burrows made by gophers. The device applies poisoned grain into the artificial burrow.

other small animals in landscape or interiorscape areas. Figure 9-24 illustrates the different types of traps available for rat and mouse control. These include snap or spring traps, glue boards, and live traps. Because they are nontoxic, traps are among the safest methods for eliminating rats, mice, and certain other

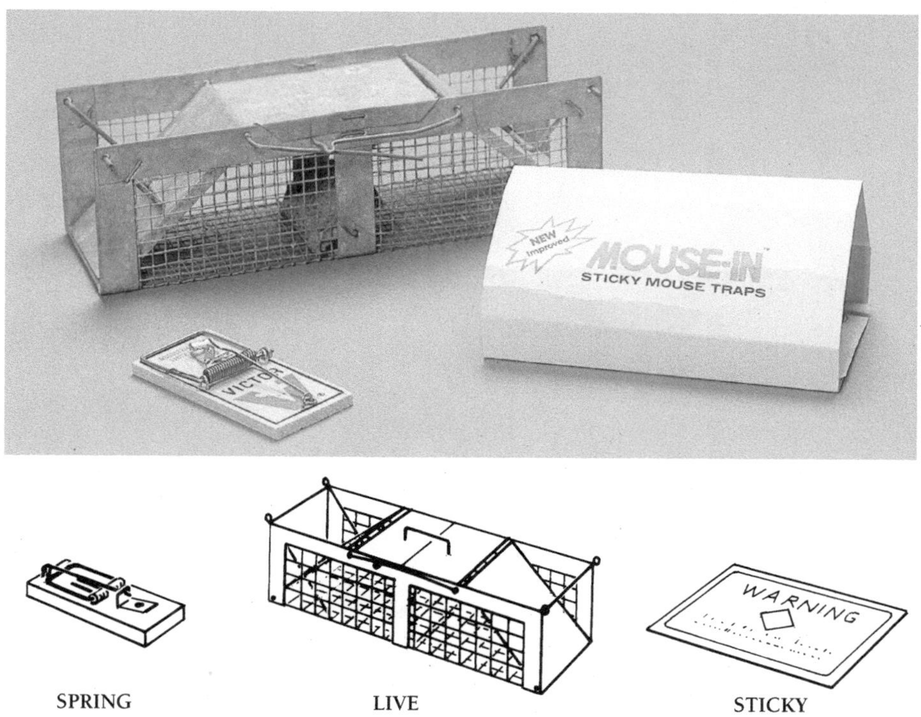

FIGURE 9-24.

Rodents may be trapped using snap or spring traps, live traps, or sticky glue boards.

small rodents. Trapping may produce quicker results than other control methods, although you may have to set many traps. You can locate and dispose of the dead animals caught in traps much more easily than those killed by poisons. Some pests can learn to avoid traps, however, and become trap-shy.

For greatest effect, select appropriately sized traps and place these along the pest's normal runways. Be sure to position the triggers of spring traps next to walls or other obstacles in the line of travel. If you use traps in an area where they are accessible to children or pets, put each inside a large, secure bait box or similar container. Trapping requires considerably more commitment of time than chemical control methods because you must check and service the traps daily.

Snap or Spring Traps. Figure 9-25 is an illustration of a typical snap trap arrangement. In many cases, snap traps can provide effective control of rodents without resorting to rodenticides. They are useful in situations where there is a chance that poisoned animals might go into inaccessible areas to die. You may also want to use snap traps where

- only a few pest individuals are present
- people in the infested area are opposed to the use of toxicants
- you can afford the investment of time that is necessary for a successful trapping program

To increase their effectiveness, place snap traps where rats or mice are most likely to encounter them, such as along or near known travel routes. Always use many traps to increase your chances of catching the target rodents. Place two traps end-to-end as shown in Figure 9-26 to prevent rodents from jumping over and avoiding the traps. Never put traps in areas that are accessible to children or pets. Also, do not put traps where someone might accidentally trip them. In interiorscape areas, put the traps out of sight so trapped animals will not be visible. Check the traps daily and remove trapped animals.

The odors from rodents that have been killed in a trap do not repel other rodents, and may even enhance the effectiveness of spring or live traps. Petroleum oils repel rodents, however, so never use these to lubricate a trap or prevent rusting. If rusting is a problem, cover the metal surfaces with lard or some other animal fat.

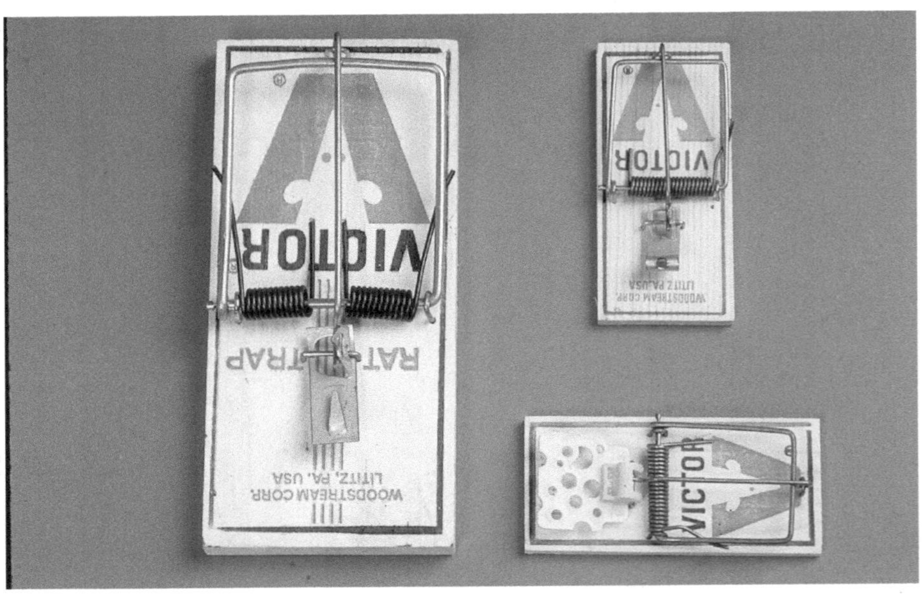

FIGURE 9-25.

Snap traps are often an effective method for controlling rats or mice, eliminating the need for using rodenticides.

FIGURE 9-26.

For greater effectiveness, place snap traps in pairs along edges or walls. This will prevent rodents from jumping over a trap to avoid being caught.

To keep pests from becoming trap shy, start out by baiting unset traps for several days until the rodents take the bait regularly. Then set the spring mechanism. Bait snap traps with an attractive food or nesting material. Foods attractive to rodents include peanut butter, bacon, sausage, whole peanuts, chocolate candy, marshmallows, cheese, and dog or cat kibble. Many people have identified other materials that work well in their traps. When unsure of which bait to use, try several different types at first, and then continue baiting with the one that appears to work best.

Tie the bait securely to the trigger mechanism so rodents cannot remove it without springing the trap. When you use baits that you cannot tie to the trap, like peanut butter, place only a small amount onto the trigger plate. The rodent must work harder to remove the bait so the chances that it will spring the trigger mechanism are increased. Replace bait frequently to keep baited traps from losing their effectiveness as the bait becomes old or stale.

Trapping rodents is most successful during the first few days after you have set out the traps. After 3 or 4 days, the rodent catch usually drops off.

Glue Boards. Glue boards, also called sticky traps, are disposable cardboard or plastic units having one or more surfaces thickly coated with a sticky paste. Use glue boards the same way as you use snap traps to catch mice and rats (Figure 9-27). Glue boards may be less effective in outdoor areas where debris and moisture can coat the sticky surfaces. In areas subject to dusty or moist conditions, place each glue board in a protected bait box and use nonpoisonous bait to attract rodents to the boxes.

Put glue boards in runways and tape them around pipes or other objects where rodents travel. Once the trap catches an animal, dispose of the trap and pest. For control of mice, glue boards have similar advantages to snap traps. Keep them away from children or pets.

Check glue board traps daily. Do not reuse a trap once you have caught a rodent in it: the trap will lose some of the sticky substance and will no longer be as effective. Rats, being larger and stronger than mice, are more difficult to capture on glue boards. If you set out glue boards for rats, secure the traps to a surface with tape or tacks to

FIGURE 9-27.

Glue boards, also known as sticky traps, are used for trapping mice, rats, and insects such as cockroaches.

keep rats from dragging them away.

Live Traps. You can also use live traps to help control rats, mice, skunks, opossums, and other small animal nuisance pests (Figure 9-28). Live trapping requires time and patience. For bait, use a nontoxic food. Suitable baits include grains, fruits, meat, and other common foods. To increase trapping success, keep traps locked in their open position and supplied with fresh bait for several days. Then set the trap closing mechanism after the rodents or other animals have learned to take the food on a regular basis. An animal that has escaped from a trap will probably not go into one again, so do a careful, thorough job of trapping the first time.

Take the appropriate measures to prevent capture of non-pest animals. For example, place live traps in areas where children or pets will not disturb them and where they will not capture cats or small dogs. Check the traps daily or more frequently to see whether you have caught anything or to make sure that bait is available and that the trap is still set.

Managing vertebrate pests requires special skills, not only because these animals are larger, but also because of the need to balance the impact of economic losses and human diseases against humane treatment of pests. One option for disposing of a live animal is to release it into another area where it will not create problems. If you cannot do this, choose a speedy and humane way to destroy the animal. Methods include putting the animal into a chamber saturated with carbon dioxide (from dry ice), shooting, and drowning.

MAINTAINING APPLICATION EQUIPMENT

Effective pesticide application depends on the proper maintenance and adjustment of application equipment. Regular programs for inspection and periodic maintenance will help you avoid accidents, leaks, and spills caused by ruptured hoses, faulty fittings, damaged tanks, or other problems.

Inspect application equipment for

FIGURE 9-28.

Live traps can be used to capture small nuisance animals, such as raccoons, without injuring them. However, care should be taken to prevent domestic animals, such as this cat, from being accidentally trapped.

wear, corrosion, or damage before each use. Replace or repair faulty components. Always keep records of maintenance performed and any repairs made to application equipment. Keep a supply of common repair parts on hand so you can quickly fix the equipment if anything goes wrong. Thoroughly clean the equipment after every application. Wear personal protective equipment, rubber gloves, and eye protection when cleaning or repairing the equipment. When you are not using your application equipment, store it in a way that prevents deterioration or damage.

Thoroughly clean dust and granule applicators after each use. Wear a dust mask while cleaning to prevent yourself from inhaling dusts. Be sure to remove all pesticide residues from the equipment. Once clean, lubricate moving parts according to the manufacturer's instructions. Inspect the equipment for wear and corrosion. Repair or replace rusted or corroded components.

APPLYING THE CORRECT AMOUNT OF PESTICIDE

There are a number of steps you should take to ensure that you apply the correct amount of pesticide in each application. Some equipment has to be calibrated before you can use it to apply pesticides to turf, landscape, and interiorscape areas so you will know how much pesticide to put into it and at what concentration. The failure to properly calibrate equipment is a main cause of ineffective pesticide applications. In addition, inaccurate calibration always carries the potential that excessive or illegal residues will remain on treated surfaces.

To calibrate your equipment, you must first determine the amount of pesticide you will want to apply: the application rate. *Check the pesticide label for this information.* Once you have calibrated your equipment, check it and test it periodically to make sure the calibration stays accurate.

Many operators fail to understand how rapidly equipment can lose calibration or become worn, and most pesticide handlers do not calibrate their application equipment often enough. Make sure that the person actually applying the pesticide for you is aware of the purpose of calibrating the equipment and understands how to spot problems that could lead to application of the wrong amount of pesticide.

Why Calibration Is Essential

The main reason for calibrating your application equipment is so you can determine just how much pesticide to put into the tank or hopper. This ensures that the correct amount of chemical will be applied. Calibration is necessary for

- effective pest control
- protection of people, the environment, and treated surfaces
- prevention of waste
- control of the volume of water applied to a given area (for liquid applications)
- compliance with the law

Effective Pest Control. Pesticide manufacturers spend millions of dollars researching ways to use their products. Their research includes determination of the correct amount of pesticide to apply to control specific pests. If you use less than the labeled amount, you may not achieve adequate control. This could result in a waste of time and money. An inadequate application of pesticide may even contribute to the development of pesticide resistance and pest resurgence. Applications in excess of label rates waste materials, and what's more they are *illegal.*

Human Health Concerns. If you apply pesticides at rates higher than are indicated on the label, you could endanger the health of people in the area. Illegal and hazardous residues may result if a pesticide is over-applied. Poorly calibrated application equipment may also expose equipment operators to concentrations of pesticide for which they are not adequately protected.

Environmental Concerns. Improper pesticide concentrations may cause environmental problems. By calibrating equipment to maintain application rates within label requirements, you help protect beneficial insects and wildlife. You also reduce the potential for contamination of surface water, groundwater, and the air.

Protecting Treated Surfaces. Certain pesticides are phytotoxic and cause damage to plant surfaces when used at above-label rates. Manufacturers evaluate this sort of potential problem when they test their chemicals to determine safe concentrations. If you use too much pesticide, you may also increase the possibility that excessive residues will build up in the soil.

Preventing Waste of Resources. Use of an improper amount of pesticide (either too much or too little) wastes time and adds unnecessary costs to the application. Not only are pesticides expensive, but the labor and the wear and tear on equipment required to make inappropriate applications are costly.

Legal Aspects. Pesticide handlers who use pesticides improperly and their employers are subject to criminal and civil charges that can result in fines, imprisonment, and civil penalties. Applicators and their employers are liable for injuries or damage caused by any pesticide application.

Equipment Calibration Methods

Liquid application equipment and granular application equipment require different types of calibration.

Calibrating Liquid Sprayers

You usually do not need to calibrate most hand-operated equipment used to apply liquid pesticides since those pesticides are mixed as percentage or parts-per-million (ppm) concentrations (see below). You simply spray the diluted solution onto plant surfaces according to label instructions. A powered liquid sprayer, however, applies pesticides on an area basis (amount per square foot or per acre) and so requires frequent calibration to compensate for wear. Abrasive pesticides such as wettable powders increase the rate of wear. Pump wear decreases the amount and pressure of fluid output. Nozzle wear increases the volume of output, which usually lowers the output pressure and may produce a poor spray pattern.

The goal of calibration is to determine how much area each tankful of spray covers when the equipment moves at a known speed and operates at a known pressure.

Small Tanks, 5 Gallons or Less. Here is a simple calibration technique for a rig equipped with small tanks (5 gallons or less):

1. Mark off an area of about 200 square feet.
2. Put a known amount of clean water into the tank and spray this over the marked-off area.
3. Measure the amount of water remaining in the tank and subtract this amount from the amount you put into the tank.
4. Divide number of gallons of liquid you used by the square footage of the area that you covered to get the volume of spray that you applied per square foot.

For example:

measured area = 200 square feet

water put into the tank = 4.5 gallons

water remaining in the tank after spraying 200 square feet = 2.0 gallons

amount sprayed on 200 square feet = 4.5 gallons – 2.0 gallons = 2.5 gallons

2.5 gallons/200 square feet = 0.0125 gallon/square foot

Repeat this process two more times. Add the three results together and then divide that sum by three to get the mean value for gallons applied per square foot. Make sure to walk at a constant speed during all applications. Now, take the total square feet of the area you intend to spray and multiply it by mean gallons per square foot to determine the total number of gallons you will need to apply.

Large Tanks, Greater Than 5 Gallons. For large turf application equipment, you need to measure these four factors:

- tank capacity
- travel speed
- flow rate
- spray swath width

Start with a clean, decontaminated sprayer and make sure to use clean water to prevent exposure of employees, yourself, or others to pesticide residues.

Tank Capacity. Physically measure the capacity of the spray tank and its associated plumbing. Never rely on the manufacturer's tank size ratings: they may be approximate volumes or they may not take into account fittings installed inside the tank. In addition, the combined capacities of any spray lines, the pump, and any filters contribute to the overall tank volume. To calibrate your equipment accurately, you need to know exactly how much liquid the sprayer holds.

Put the sprayer on a perfectly level surface. Make sure the tank is completely empty and close all valves to prevent water leaks. Add measured amounts of clean water until you completely fill the tank, and make sure the pump and hoses are filled, too. To measure the water as you fill the tank, use either a flow meter attached to a hose or a bucket or other container of known volume. A 5-gallon bucket works well for smaller spray tanks. Calibrate and mark the bucket before using it to fill the tank. If you are not using a flow meter, use smaller-volume calibrated containers to top off the tank. Record the total volume of water that you put into the tank and then paint or engrave this figure onto the outside of the tank for permanent reference.

While you are filling the tank, calibrate the tank's sight gauge. Make marks on the tank or the gauge as you add measured volumes of water. If the unit does not have a sight gauge, mark volume increments onto a dipstick that you will always afterward keep with the tank. Mark the tank, gauge, or dipstick in 1-gallon increments for tanks with a capacity of 10 gallons or less. Use increments of 5 or 10 gallons for tanks with a total capacity of 50 gallons or less. On tanks larger than this, use increments of 10 to 20 gallons. Once you calibrate the sight gauge, tank, or dipstick, you can measure how much liquid is in the tank even when it is not entirely full. Always return the tank to a level surface before you read the sight gauge or dipstick.

Travel Speed. Measure travel speed under actual working conditions. For instance, if you are calibrating a turf sprayer, take the water-filled tank to the turf area. When calibrating a backpack or hand-held sprayer, walk on terrain that is similar to the area that you plan to spray.

Using a 100-foot tape, measure off any convenient distance. It can be more or less than 100 feet, but the accuracy of your calibration increases as you measure it over longer distances (between 200 and 300 feet). Sometimes a pesticide applicator will calibrate his or her equipment based on multiples of 88 feet, since 88 feet is the distance covered in 1 minute by a vehicle traveling 1 mile per hour. Indicate the beginning and end of the measured distance with colored flagging tape wrapped around small stakes.

Have someone drive (or walk, if calibrating a backpack sprayer) the sprayer through the measured distance. Maintain the speed desired for

SIDEBAR 19

Calculating Application Travel Speed

1. Convert minutes and seconds into minutes by dividing the seconds (and any fraction of a second) by 60.

EXAMPLE: Your trip took 1 min and 47.5 sec

$$\frac{47.5 \text{ sec}}{60 \text{ sec/min}} = 0.79 \text{ min}$$

... add the two amounts together:

1 min + 0.79 min = 1.79 min

2. Add the *converted* minutes from each run and divide by the number of runs.

EXAMPLE: Three runs were made. . .

Run #1 = 1 min, 47.5 sec = 1.79 min

Run #2 = 1 min, 39.8 sec = 1.66 min

Run #3 = 1 min, 52.0 sec = 1.87 min

Total = 5.32 min

$$\frac{5.32 \text{ min}}{3 \text{ runs}} = 1.77 \text{ min/run average time}$$

3. Divide the measured distance (in feet) by the average time. This will tell you how many feet were traveled in one minute.

EXAMPLE: The measured distance for this example is 187 feet.

$$\frac{187 \text{ ft}}{1.77 \text{ min}} = 105.65 \text{ ft/min}$$

4. If you wish to determine the speed in miles per hour, divide the feet-per-minute figure by 88 (the number of feet traveled in 1 minute at 1 mile per hour).

EXAMPLE:

$$\frac{105.65 \text{ ft/min}}{88 \text{ ft/min/mi/hr}} = 1.20 \text{ mi/hr}$$

an actual application. Choose a speed within a range that is appropriate for the application equipment. Make sure that the equipment or person gets up to the actual application speed before crossing the first (starting) marker. Use a stopwatch to determine how long, in minutes and seconds, it takes to travel the measured distance. For best results, repeat this process 2 or 3 times and take a mean of the results. Follow the procedure in Sidebar 19 to calculate the actual speed of the equipment.

Flow Rate. Measure the actual output of the sprayer when nozzles are new, and then re-measure it periodically thereafter so you can allow for nozzle wear. Manufacturers provide charts that show the initial output of specific nozzle sizes at specified sprayer pressures, but you should check the output yourself under actual conditions of operation. Manufacturers' charts are most accurate when you are using new nozzles. Nozzles that have been in use for a period of time may have a different output rate because of wear. Even new nozzles may have slight variations in their actual output. Sprayer pressure gauges may also be inaccurate, further adding error to any output estimate that is based on manufacturers' charts.

You measure liquid sprayer output (the flow rate) in gallons per minute. To determine the flow rate, collect the volume of water discharged out of the nozzle over a measured period of time. If the sprayer has more than one nozzle, collect liquid from each nozzle separately; that way, you can compare the output from each nozzle and find any malfunctions or wear. You will need a stopwatch and a calibrated container with which to make measurements. Wear rubber gloves to avoid skin contact with the liquid and eye protection to prevent you from getting spray droplets into your eyes. Position yourself upwind from the nozzles to prevent any fine mist or spray from contacting your face and clothing.

Fill the tank at least half full with water. If it is a powered unit, start the sprayer and bring the system up to normal operating pressure. Most power sprayers have a limited operating pressure range depending on the type of pump and type of power unit. Never operate equipment beyond its normal working range, since this may damage the pump. Adjust the pressure to the requirements of the spray situation and the nozzle manufacturer's recommendations.

While all nozzles are operating at the proper pressure, collect about 15 to 30 fluid ounces of liquid from each nozzle (Figure 9-29). Use a stopwatch to determine the number of seconds it takes to collect each volume.

If you are calibrating a backpack sprayer, pump the unit as you would during an actual application. Collect spray into a calibrated container for a measured period of time. If you are calibrating a compressed-air sprayer, remember that it will lose pressure during operation, so you must frequently pump it up. The best way to calibrate a compressed-air sprayer is to fill the tank about one-half full with water. This provides a sufficient volume of air to maintain a uniform pressure during the calibration process.

Record the volume of liquid collected from each nozzle and the number of seconds it takes you to collect each amount. Use a format similar to that shown in Figure 9-30. Determine the fluid-ounces-per-second output for each nozzle by dividing the volume by the seconds required to collect it (see Sidebar 20). Convert the fluid-ounces-per-second into gallons-per-minute by multiplying the result by the constant 0.4688. This constant represents 60 seconds per minute divided by 128 fluid ounces per gallon.

If you are calibrating a sprayer that has several nozzles, the output will usually vary from one nozzle to another. In the example in Sidebar 20, part 1, the output ranges from 0.250 gallons per minute up to 0.328 gallons per minute. If we assume that the rated capacity (as given by the manufacturer) for these nozzles at the recommended operating pressure is 0.250 gallons per minute, then the variation among nozzles should not be greater than 5% and the output of any one nozzle should not exceed the manufacturer's rated output by more than 10%. Figure the percentage of variation (example in Sidebar 20, part 2) by dividing the actual output by the rated output. Subtract 1.00 from this figure and then multiply the result by 100 to obtain the percentage of variation. Nozzles 3 and 5 in the Sidebar 20 example exceed these amounts and must therefore be replaced. It is important to remember

FIGURE 9-29.

To determine the output from each nozzle, collect water over a measured period. Make sure the sprayer is operating at the pressure that would be used under actual field conditions. Wear rubber gloves and eye protection since the water may contain traces of pesticide.

that whenever you replace any nozzles you need to re-check the flow rate of all nozzles, because changing one nozzle may affect the pressure in the whole system. After you change nozzles, re-adjust the pressure regulator to maintain the desired pressure. Sidebar 21 shows how to recalculate the gallons-per-minute output after you replace worn nozzles.

Swath Width. The final measurement needed to complete calibration is the width of the spray swath applied by the sprayer. Figure 9-31 illustrates spray swath widths for various application situations. For multiple-nozzle boom sprayers, the swath is the width of the boom plus the distance between one pair of nozzles. You can also calculate swath width by multiplying the num-

NOZZLE NO.	VOLUME OUTPUT (FL. OZ.)	TIME (SECONDS)
1	12.5	23.2
2	12.0	22.5
3	15.5	24.8
4	14.5	26.1
5	19.0	27.2
6	13.0	23.9

FIGURE 9-30.

Sample data sheet for recording output from nozzles. This data is used in the example in Sidebar 20.

SIDEBAR 20

Calculating Gallons per Minute for Low-Pressure Sprayers

1. Determine the gallons-per-minute output of each nozzle by dividing the fluid ounces collected by the time (in seconds) and multiplying the result by 0.4688.

EXAMPLE:

NOZZLE NO.	FL. OZ./SEC.			x	0.4688	=	GPM
1	12.5/23.2	=	0.539	x	0.4688	=	0.253
2	12.0/22.5	=	0.533	x	0.4688	=	0.250
3	15.5/24.8	=	0.625	x	0.4688	=	0.293
4	14.5/26.1	=	0.556	x	0.4688	=	0.261
5	19.0/27.2	=	0.699	x	0.4688	=	0.328
6	13.0/23.9	=	0.544	x	0.4688	=	0.255
Total output = 1.640 gpm							

2. Compute the percentage variation from the rated nozzle output by dividing the actual gallons-per-minute output by the rated output. Subtract 1 from the result and multiply by 100.

EXAMPLE:

NOZZLE NO.	ACTUAL GPM/RATED GPM	SUBTRACT 1.00		MULTIPLY BY 100	=	PERCENT VARIATION
1	0.253/0.250	= 1.012 – 1.00	=	0.012 x100	=	1.2
2	0.250/0.250	= 1.000 – 1.00	=	0.000 x100	=	0.0
3	0.293/0.250	= 1.172 – 1.00	=	0.172 x100	=	17.2*
4	0.261/0.250	= 1.044 – 1.00	=	0.044 x100	=	4.4
5	0.328/0.250	= 1.312 – 1.00	=	0.312 x100	=	31.2*
6	0.255/0.250	= 1.020 – 1.00	=	0.020 x100	=	2.0

Nozzles 3 and 5 must be replaced because they exceed the manufacturer's rated output by more than 50 percent.

ber of nozzles by the nozzle spacing (number of nozzles X nozzle spacing = swath width). When making a pesticide application with a boom sprayer, overlap the spray from two passes by the same distance as the overlap of two adjacent nozzles on the boom (Figure 9-32). Adjust the boom height so that there is approximately a 30% overlap of spray from adjacent nozzles on the boom (Figure 9-33). Position nozzles at the exact height they would be during an actual application. Check the spray boom to make sure it is level (an unleveled boom causes uneven spray distribution [Figure 9-34]). Align fan nozzles as illustrated in Figure 9-35 to give an even spray distribution.

Measure the swath width of a backpack sprayer based on the spray pattern left on the ground after a test run. Keep the nozzle at the height you would use during an actual application and maintain this height always to prevent any variation in swath width. Overlap swaths only enough to ensure a uniform application pattern.

Determining the Amount of Pesticide to Use. Use tank volume, travel speed, flow rate, and swath width to calculate the total area covered with each tankful of material. Knowing this value, you will be able to determine how much pesticide to put into the tank. Choose from the two calculation methods described here. One applies to pesticides applied

SIDEBAR 21

Recalculating Output after Replacing Worn Nozzles

1. Worn nozzles (numbers 3 and 5 in Sidebar 20) have been replaced. Now re-measure the output of all nozzles on the boom. Recalculate the gallons-per-minute for each nozzle. Add these rates together to determine the total output of the sprayer.

EXAMPLE:

NOZZLE NO.	FL. OZ./SEC.			x	0.4688	=	GPM
1	12.5/23.2	=	0.539	x	0.4688	=	0.253
2	12.0/22.5	=	0.533	x	0.4688	=	0.250
3	13.3/24.8	=	0.542	x	0.4688	=	0.254
4	14.5/26.1	=	0.556	x	0.4688	=	0.261
5	15.2/28.3	=	0.537	x	0.4688	=	0.252
6	13.0/23.9	=	0.544	x	0.4688	=	0.255
Total output = 1.525 gpm							

2. Check to see that all nozzles are within 5 percent of the rated capacity of the nozzles.

EXAMPLE:

NOZZLE NO.	ACTUAL GPM/RATED GPM	SUBTRACT 1.00		MULTIPLY BY 100	=	PERCENT VARIATION
1	0.253/0.250	= 1.012 - 1.00	=	0.012 x100	=	1.2
2	0.250/0.250	= 1.000 - 1.00	=	0.000 x100	=	0.0
3	0.254/0.250	= 1.016 - 1.00	=	0.016 x100	=	1.6
4	0.261/0.250	= 1.044 - 1.00	=	0.044 x100	=	4.4
5	0.252/0.250	= 1.008 - 1.00	=	0.008 x100	=	0.8
6	0.255/0.250	= 1.020 - 1.00	=	0.020 x100	=	2.0

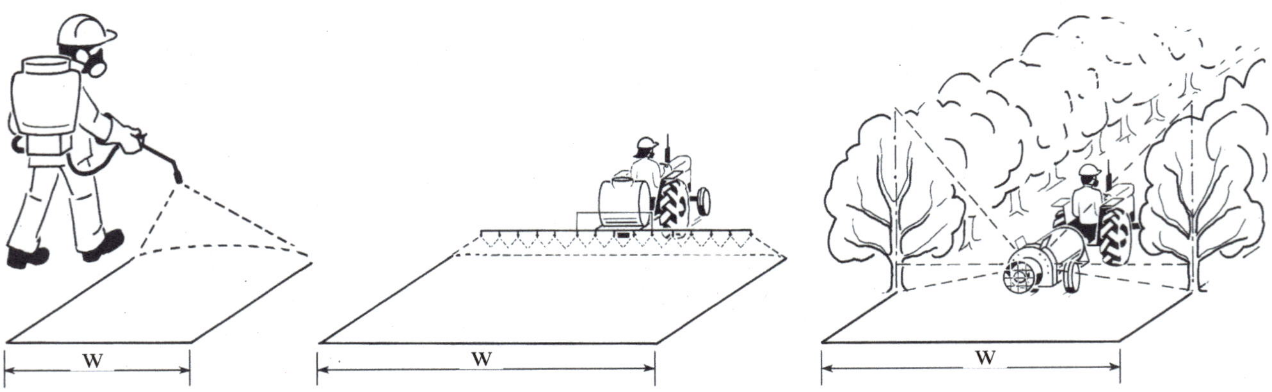

FIGURE 9-31.

A spray swath is the horizontal width being covered with spray material during a single pass of the applicator.

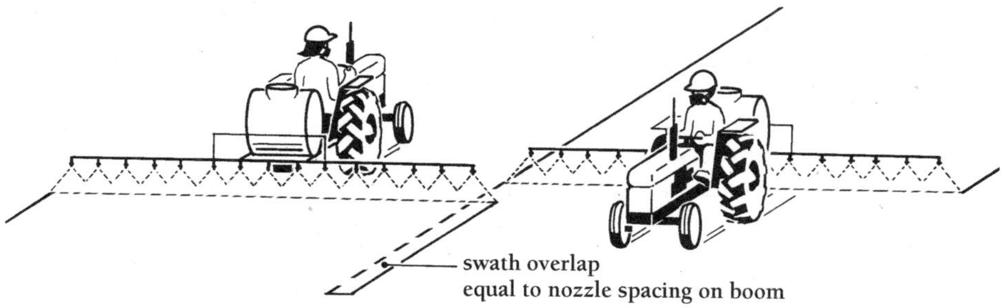

FIGURE 9-32.

Spray from adjacent swaths should overlap by the same amount as the spray from adjacent nozzles on the spray boom overlaps (usually about 30% of the spray pattern of one nozzle). To achieve this degree of overlap, allow one nozzle-width spacing between swaths.

by the acre (Sidebar 22), which may be useful for large turf areas. The other is for applications made by the square foot (Sidebar 23), and is appropriate for small lawn or other landscape treatments. To prevent waste of pesticide material, accurately measure the area that you plan to treat. Then mix only the amount of pesticide needed for that application.

Calibrating Granule Applicators

The technique for calibrating granule applicators is similar in many ways to that used for liquids. The difference is that you have to calibrate a granule applicator for each type of granular pesticide that you apply. You should also recalibrate the equipment each time weather conditions change, since temperature and humidity influence the flow of granules. In addition, granules vary in size and shape from one pesticide to the next, and this also affects their flow rates.

Before you begin to calibrate a granule applicator, make sure that it is clean and that all parts are working properly. Most equipment requires regular lubrication, so check to see if the equipment's moving parts run smoothly.

Always wear rubber gloves to prevent contact with any pesticide residues

FIGURE 9-33.

Under normal conditions, flat fan nozzles on a spray boom must be spaced so there is a 30% overlap of the spray emitted by adjacent nozzles. This provides a uniform distribution of spray.

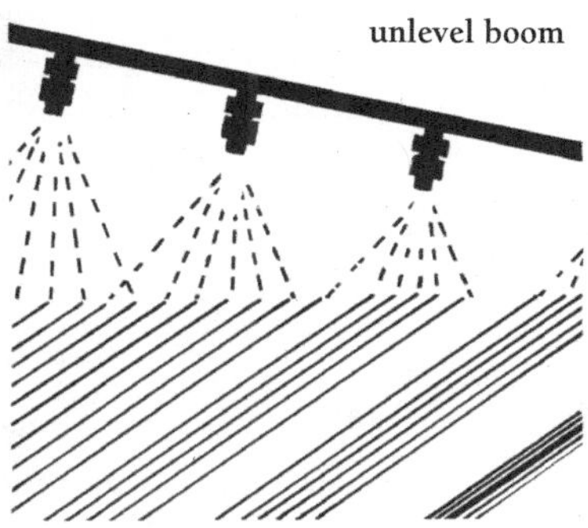

FIGURE 9-34.

An unleveled spray boom will cause an uneven pesticide application.

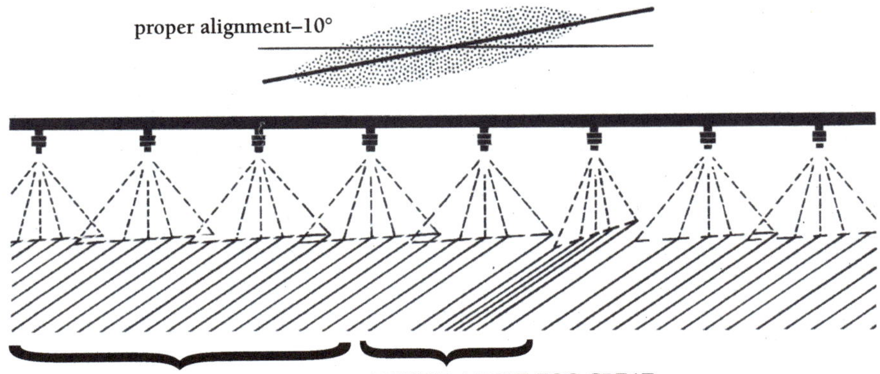

FIGURE 9-35.

The spray pattern will be uneven if nozzles are not aligned properly on the spray boom. Rotate nozzles about ten degrees from the axis of the boom to prevent droplets from adjacent nozzles from touching, but still allow for proper overlap of the spray pattern.

SIDEBAR 22

How Much Pesticide to Put into the Spray Tank (pesticides applied on a per-acre basis)

1. First, determine how large an area you can treat in 1 minute. Divide the spray swath width by 43,560 (the number of square feet in 1 acre) and multiply the result by the sprayer's travel speed in feet per minute. The result will be the acres treated per minute. In the example in Sidebar 19, travel speed was calculated as 105.65 feet per minute. Assuming that the swath width is 12 feet, the calculation would be

EXAMPLE:

$$\frac{12 \text{ ft}}{43560 \text{ sq ft/ac}} \times 105.65 \text{ ft/min} = 0.0291 \text{ ac/min}$$

In this example, when a swath 12 feet wide is being sprayed, 0.0291 acres are covered in 1 minute.

2. Next, determine the gallons of liquid being applied per acre. Divide the gallons-per-minute figure by the acres per minute.

EXAMPLE:

$$\frac{1.525 \text{ gal/min}}{0.0291 \text{ ac/min}} = 52.4 \text{ gal/ac}$$

3. Then determine the number of acres that can be treated with a full tank. Divide the actual measured volume of the spray tank (or tanks) by the gallons-per-acre figure. For the purposes of this example, assume the tank holds 55.5 gallons when filled.

EXAMPLE:

$$\frac{55.5 \text{ gal/tank}}{52.42 \text{ gal/ac}} = 1.06 \text{ ac/tank}$$

4. Finally, determine how much pesticide to put in the tank. Multiply the acres per tank by the pesticide's recommended per-acre rate (check the pesticide label for this information).

EXAMPLE:

PESTICIDE LABEL SAYS		ACRES PER TANK		AMOUNT OF PESTICIDE TO PUT IN TANK
1.5 lb/ac	x	1.06	=	1.59 lb
3 qt/ac	x	1.06	=	3.18 qt
2 gal/ac	x	1.06	=	2.12 gal
1 pt/ac	x	1.06	=	1.06 pt

SIDEBAR 23

How Much Pesticide to Put into the Spray Tank (pesticides applied by the square foot)

1. Determine how many square feet can be treated in 1 minute. Multiply the sprayer's ground speed (as determined by the procedures in Sidebar 19) by its swath width. In this example, assume that a single-nozzle hand-operated sprayer is being used to apply a swath width of 2.5 feet at a speed of 105.65 feet per minute.

EXAMPLE:

$$105.65 \text{ ft/min} \times 2.5 \text{ ft} = 264.13 \text{ sq ft/min}$$

2. Next, determine the volume of spray, in gallons, that will be applied to 1 square foot. Divide the sprayer's gallon-per-minute output (see Sidebar 21) by the square feet per minute. For this example, assume that the backpack unit sprays 0.05 gallons per minute.

EXAMPLE:

$$\frac{0.05 \text{ gal/min}}{264.13 \text{ sq ft/min}} = 0.00019 \text{ gal/sq ft}$$

3. Then find out how many square feet can be sprayed with one tank. Divide the gallons per square foot into the measured tank capacity. For this example, assume that the tank holds 3 gallons.

EXAMPLE:

$$\frac{3 \text{ gal/tank}}{0.00019 \text{ gal/sq ft}} = 15{,}789 \text{ sq ft/tank}$$

4. Finally, determine how much pesticide to put in the tank. First, read the pesticide label; it will tell you the amount of pesticide to apply. Normally, the label will tell you how much to apply per square foot (or per 100 or 1000 square feet) or per acre.

EXAMPLE:

If the label gives the dosage rate per 1, 100, or 1000 square feet, multiply that rate by the square-feet-per-tank as determined in step 3:

PESTICIDE LABEL SAYS	x	SQUARE FEET PER TANK	=	AMOUNT OF PESTICIDE TO PUT IN TANK
$\frac{3 \text{ fl oz}}{1000 \text{ sq ft}}$	x	15,789	=	47.37 fl oz
$\frac{3/4^{*} \text{ fl oz}}{1000 \text{ sq ft}}$	x	15,789	=	11.84 fl oz
$\frac{1 \text{ oz}}{100 \text{ sq ft}}$	x	15,789	=	157.89 oz

*The fraction ¾ is converted to its decimal equivalent 0.75 to complete this calculation.

EXAMPLE:

If the pesticide label gives the dosage rate in units of pesticide per acre, convert square feet per tank (from step 3) to acres per tank by dividing it by 43,560 (there are 43,560 square feet in one acre):

$$\frac{15{,}789 \text{ sq ft/tank}}{43{,}560 \text{ sq ft/ac}} = 0.362 \text{ ac/tank}$$

Then multiply the labeled rate per acre by the acres-per-tank figure:

PESTICIDE LABEL SAYS	X	ACRES PER TANK	=	AMOUNT OF PESTICIDE TO PUT IN TANK
1.5 lb/ac	x	0.362	=	0.543 lb (8.69 oz)
3 qt/ac	x	0.362	=	1.09 qt (34.6 fl oz)
2 gal/ac	x	0.362	=	0.724 gal (5.8 pt)
1 pt/ac	x	0.362	=	0.362 pt (5.8 fl oz)

that may remain on the equipment. Calibrating a granule applicator involves using actual pesticides, so wear the label- and regulation-prescribed personal protective equipment. Some formulations are dusty and may require respiratory protection.

You must measure three variables when calibrating a granule applicator:

- travel speed
- output rate
- swath width

Travel Speed. Determine the unit's travel speed in feet-per-minute in the same way as for liquid applicators. Follow the instructions given earlier in this chapter. Fill the applicator's hopper with the pesticide granules so you can measure its speed under actual operating conditions.

Output Rate. To determine the rate of output, fill the hopper with the granular pesticide. Set the adjustable opening on the granule hopper according to the manufacturer's instructions for the approximate application rate and speed that you need. Once you set up the equipment, use one of the following methods to determine the actual output rate:

1. *Measure the Quantity of Granules Applied to a Known Area.* Collecting and weighing the granules applied to a known area is the easiest way to calibrate a granule applicator. Spread out a plastic tarp of known size (e.g. 10 feet by 10 feet) on the ground. Then operate the broadcast applicator at a known speed across the tarp (Figure 9-36). Place the granules collected by the tarp into a container and weigh them.

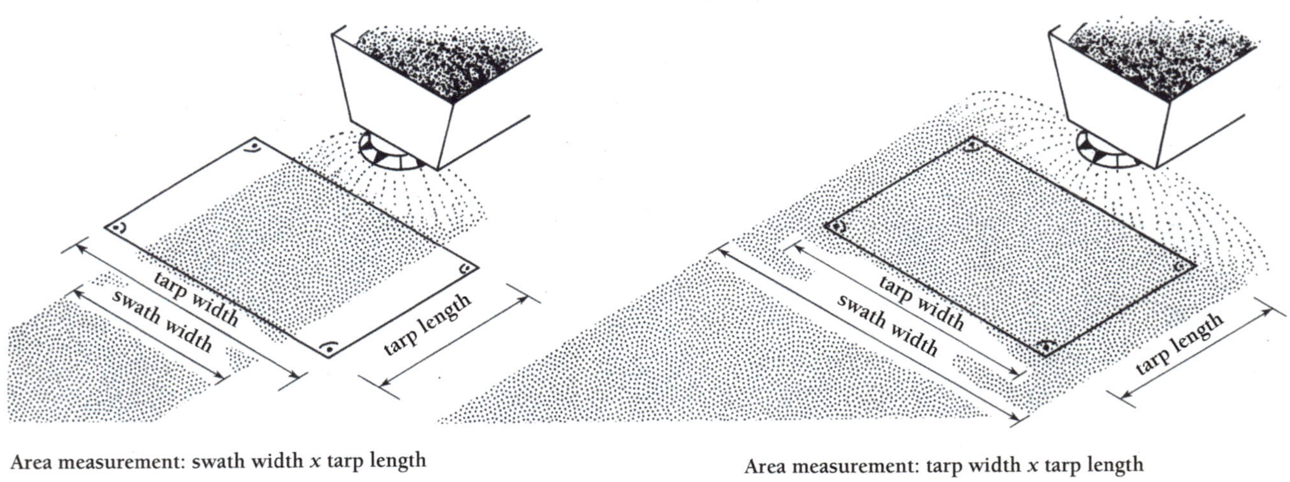

FIGURE 9-36.

To determine the actual application area for granules, measure the swath width across a plastic tarp and multiply this by the length of the tarp. If the swath is wider than the tarp, compute the area by multiplying the length by the width of the tarp.

SIDEBAR 24

Calculating Granule Output Rate by Measuring the Quantity Applied to a Known Area

1. Spread a plastic tarp on the ground and measure its length. In this example, the length is 10 feet.
2. Fill the hopper of the granule applicator, adjust the output port to the recommended opening, and travel across the tarp at a known speed while granules are being applied.
3. Measure the swath width of the granules that were applied and compute the area of the swath. In this example, the swath width is 3 feet, so the area treated is 3 feet multiplied by 10 feet (30 square feet).
4. Transfer all the granules on the tarp into a container and weigh them. In this example, the granules weigh 18 ounces. Convert this to pounds by dividing the weight of the granules by 16 ounces per pound.

EXAMPLE:

$$\frac{\text{18 ounces}}{\text{16 oz/lb}} = \text{1.125 lb}$$

5. Divide the weight of the granules collected (in pounds) by the square feet of the swath.

EXAMPLE:

$$\frac{\text{1.125 lb}}{\text{30 sq ft}} = \text{0.0375 lb/sq ft}$$

In this illustration, the granule applicator is broadcasting 0.0375 pounds of material per square foot. If the label calls for a greater amount, open the port more or slow the applicator's speed of travel. If the label calls for a lesser amount, close the port some more or speed up the rate of travel. After you have made an adjustment, repeat this calibration procedure.

Use the calculations in Sidebar 24 to figure the amount of granular pesticide applied per square foot.

OR

2. *Refill the Hopper After a Measured Amount of Time.* Use this method with hand-operated equipment or when you will be applying small quantities. Fill the hopper to a known level and operate the equipment for a measured period of time. When the time is up, weigh the quantity of granules required to refill the hopper to its original level. Use the calculations in Sidebar 25 to compute the output rate. Any settling of granules in the hopper may cause this method to be less accurate than method 1 listed above.

Swath Width. To measure the swath width, operate the equipment under actual field conditions. Whenever possible, place cans, trays, or other containers at even intervals across the application swath. Use these to collect granules. Weigh the granules collected in each container separately to determine the distribution pattern. You can operate some spreaders over a strip of black cloth or black plastic. This gives you a quick visual estimate of granule distribution and swath width.

SIDEBAR 25

Calculating the Rate of Output by Refilling the Hopper after a Measured Period of Time

1. Fill the hopper to a known level with granules.
2. Operate the equipment for a measured period of time at a known speed.
3. Weigh the amount of granules required to refill the hopper to its original level.

EXAMPLE:

In this example, 6.2 pounds of granules are applied in 2.5 minutes.

4. Convert the output to pounds per minute by dividing the weight of the granules during the measured time.

EXAMPLE:

$$\frac{6.2 \text{ lbs}}{2.5 \text{ min}} = 2.48 \text{ lb/min}$$

5. Use the technique shown in Sidebar 27 to calculate the rate per unit of area.

SIDEBAR 26

Calculating Granule Application Rate per Unit of Area

The area to which you will be applying granules is 1200 square feet. You will be using a drop spreader that has a swath width of 3 feet.

1. To determine the square feet being treated per minute, divide the speed of travel by the swath width. In this example, the swath is 3 feet wide and the application speed is 200 feet per minute (2.27 miles per hour).

EXAMPLE:

$$\frac{200 \text{ ft/minute}}{3 \text{ feet}} = 66.67 \text{ sq feet/min}$$

2. To determine how many pounds of formulated pesticide you are applying per square foot, divide the output rate of the granule applicator (as computed from the calculations given in Sidebar 26) by the sq feet/min calculated in step 1 of this sidebar. This example uses 2.48 pounds per minute as the output rate.

EXAMPLE:

$$\frac{2.48 \text{ lb/min}}{66.67 \text{ sq ft/min}} = 0.037 \text{ lb/sq foot}$$

3. To determine how many pounds of granules you will apply to get proper coverage of 1200 square feet, multiply the total square feet by the 0.037 lb/sq ft.

EXAMPLE:

$$1200 \text{ sq ft} \times 0.037 \text{ lb/sq ft} = 44.4 \text{ lbs}$$

Application Rate. Follow the instructions in Sidebar 26 to calculate the actual rate of granules applied per unit of area. If your calculations do not correspond to the labeled rate, adjust the equipment and repeat the calibration procedure. Motorized and hand-operated applicators apply granules at a fixed output rate, independent of ground speed. When ground speed increases, you end up applying fewer granules per unit of area. When ground speed decreases, you apply more material. With this type of equipment, you can adjust the application rate either by adjusting the size of the port opening or by changing the speed of travel.

The output of certain ground-wheel-driven granule applicators varies with the unit's ground speed. These applicators are equipped with an auger or chain that feeds the granules onto the spinning disc. With this type of equipment, if ground speed increases, the feed mechanism will run faster and the output rate will increase. When the ground speed slows down, the output rate decreases because the feed mechanism runs more slowly. The result of this automatic change in output is that the equipment applies nearly the same amount of material per unit of area regardless of its speed of travel. The equipment has minimum and maximum operating speeds determined by the manufacturer, however.

Calculations for Percentage Solutions and Parts-per-Million Solutions

Not all pesticide application rates are given in liquid volume of pesticide per unit of area. Some labels require that pesticides be mixed as a percentage solution or be diluted to a specific number of parts per million (ppm). Before you add a pesticide product to the spray tank, make sure to read and clearly understand the dilution instructions on the label.

Percentage Solutions

Sometimes label directions require that you mix the pesticide as a percentage solution. When you do this, you mix the active ingredient (a.i.) so as to achieve a known concentration, regardless of the sprayer output rate. Percentage solutions are mixed on a weight/weight basis (w/w), meaning pounds of a.i. per pound of water (one gallon of water weighs 8.345 pounds). Sidebar 27 provides an example of calculating a percentage solution with liquid formulations.

Parts-per-Million Solutions

Other pesticides must be mixed to parts-per-million (ppm) concentrations. These are calculated the same way as percentage solutions, but the ratio of active ingredient to water is much smaller. For example, a 100-ppm solution is equivalent to a 0.01% solution (Table 9-5). The ppm designation

TABLE 9-5.

Equivalent Values for Parts per Million (ppm).

PPM	DECIMAL	PERCENT SOLUTION
1 ppm	0.000001	0.0001%
10 ppm	0.00001	0.001%
100 ppm	0.0001	0.01%
1,000 ppm	0.001	0.1%
10,000 ppm	0.01	1.0%
100,000 ppm	0.1	10.0%
1,000,000 ppm	1.0	100.0%

SIDEBAR 27

Percentage Solutions—Liquid Formulations

To prepare a percent solution using a liquid formulation, you need to know the volume of the spray tank, the weight of active ingredient per gallon of formulation, and the weight of a gallon of water. The weight of water is a constant, being approximately 8.34 pounds. Assume that you have measured the volume of the spray tank and find that it holds 14.5 gallons of water. You are preparing a 1% solution of glyphosate for control of weeds, to be sprayed from a trailer-mounted sprayer with a hand-held spray nozzle. The formulation of glyphosate that you are to use contains 5.4 pounds of active ingredient per gallon.

1. Find the total weight of the liquid in the filled tank by multiplying 14.5 gallons by 8.34 pounds per gallon:

EXAMPLE:

$$14.5 \text{ gal} \times 8.34 \text{ lb/gal} = 120.93 \text{ lb}$$

2. Multiply this weight by 0.01 (1%) to determine the weight of a.i. required to mix a 1% solution:

EXAMPLE:

$$120.93 \times 0.01 = 1.21 \text{ lb}$$

3. Divide the required weight of a.i. by the weight of a.i. in the formulation. The result is the number of gallons of liquid formulation that should be added to 14.5 gallons of water to achieve a 1% solution:

EXAMPLE:

$$\frac{1.21 \text{ lb a.i.}}{5.4 \text{ lb a.i./gal}} = 0.22 \text{ gal formulation}$$

In this example, one tank of liquid should contain 0.22 gallons of glyphosate formulation. The total volume of water combined with the glyphosate formulation should equal 14.5 gallons, the capacity of the tank. You would therefore use 14.28 gallons of water and 0.22 gallons of formulated glyphosate.

NOTE: These calculations give a close approximation of the amount of liquid formulation that you would add to the tank to achieve a solution of a known percentage. The mathematics for a more exact figure are more complex, but are unnecessary for this type of work.

represents the parts of active ingredient of pesticide per million parts of water. Parts-per-million is a common way of measuring highly diluted concentrations of pesticides. Use the formulae in Sidebar 28 when calculating ppm.

Determining Application Rates for Soil Injection and Soil Drenching

Follow the product label directions to calculate the application rate for a soil injection or a soil drench application. It is important that you accurately calculate the size of your treatment site. Product labels usually tell you the amount of pesticide needed per inch of plant height or per inch of trunk diameter at breast height (DBH; about 4 to 5 feet from the ground). For grouped plantings, calculate the application rate based on cumulative feet of height (Figure 9-37). To find the cumulative feet of height, determine the height of each tree or shrub in the grouping and then add all the heights together (Sidebar 29).

When treating individual trees or shrubs, you can calculate the application rate using the plant's height or its DBH. For example, suppose that a label

SIDEBAR 28

Calculating Parts-per-Million Dilutions for Liquid Formulations

Assume that a pesticide contains 5.4 pounds of a.i. in 1 gallon of formulation. You are required to prepare a 100 ppm concentration in a 25-gallon tank.

1. Find the total weight of the liquid in the filled tank by multiplying 25 gallons by 8.34 pounds per gallon.

EXAMPLE:

25 gal/tank x 8.34 lb/gal = 208.5 lb/tank

2. Determine how many pounds of a.i. are required for a pound of spray solution.

EXAMPLE:

$$\frac{\text{100 parts a.i}}{\text{1,000,000 parts sol}} = 0.0001$$

You will need 0.0001 pounds of a.i. for each pound of solution in order to achieve a 100 ppm mixture.

3. Determine how many pounds of a.i. are required for a tank of solution, using the weight of the liquid in the tank.

EXAMPLE:

208.5 lb/tank x 0.0001 lb a.i. = 0.021 lb a.i./tank

4. Divide the required weight of a.i. by the pounds of a.i. per gallon to determine how many gallons of formulation you will require. Since this will probably be a small number, multiply by the number of fluid ounces in a gallon (128) to convert to fluid ounces.

EXAMPLE:

$$\frac{\text{0.021 lb a.i./tank}}{\text{5.4 lb a.i./gal}} = \text{0.00386 gal/tank}$$

0.00386 gal/tank x 128 fl oz/gal = 0.49 fl oz/tank

By adding 0.49 fluid ounces of this formulated pesticide to 25 gallons of water, you will get a 100 ppm solution.

calls for an application of 1.2 teaspoons (tsp) of pesticide for every inch of trunk diameter. You plan to soil drench a tree with a 15-inch DBH. To determine the application rate, multiply the DBH by the application rate:

15 inches × 1.2 tsp = 18 tsp

(there are 6 teaspoons in a fluid ounce, so 18 teaspoons are equal to 3 fluid ounces)

18 tsp ÷ 6 tsp/fl oz = 3 fl oz

From this calculation, you know you need to apply a total of 3 fluid ounces of the pesticide product to the soil surrounding the tree.

As with most pesticide applications, the pesticide used in a soil drench or soil injection application must be diluted with water. For a soil drench application, the label may tell you to dilute the pesticide in at least 10 gallons of water per 1,000 square feet

FIGURE 9-37.

*For grouped plantings, calculate the application rate based on **cumulative feet** of height.*

SIDEBAR 29

Calculating Application Rates Based on Cumulative Feet of Height

Insecticide will be applied to a bed of 5 shrubs as a soil drench. The labeled rate is 1 ounce of product per 20 *cumulative feet* of height (Figure 9-37). To calculate the amount of insecticide product needed:

1. Measure the height of each shrub. Note that shrubs A and B are the same heights, as are shrubs, C, E, and F. You can reuse a single measurement when each of several shrubs are about the same height.
2. Add together the heights of all shrubs.
3. This sum is the *cumulative feet* of height.

SHRUB	HEIGHT IN FT.
A	6
B	6
C	2
D	12
E	2
F	2
cumulative feet	30

4. Divide this cumulative feet by the rate to determine the amount of insecticide to apply:

30 cumulative feet ÷ 20 cumulative feet per oz. of insecticide = 1½ oz. of insecticide

Additionally, for a soil drench application, the label says to dilute the product in at least 10 gallons of water per 1,000 square feet of treated surface. The 6 shrubs are growing in a 4 ft. by 25 ft. soil bed. Therefore, the treated surface will be 100 sq. ft. (4 ft. x 25 ft.=100 ft.). The 11⁄2 oz. of pesticide should be diluted in at least 1 gallon of water, then applied to soil. See the section "Determining Application Rates for Soil Injection and Soil Drenching," for an explanation of this calculation.

of treated surface. Whether you are treating a single plant or several shrubs or trees grouped into a planting, you will need to determine the number of square feet you intend to treat and then calculate the dilution rate. Using the above example, assume you are treating a single tree and you want to apply a soil drench around the base of the tree to cover an area of 10 feet by 10 feet. Multiply these lengths together to find the number of square feet you will be treating:

10 feet × 10 feet = 100 sq ft

Determine how much water to apply per square foot by dividing the 10 gallons per 1,000 square feet figure by 1,000 square feet:

10 gallons ÷ 1,000 sq ft = 0.01 gallon/sq ft

Multiply that number by the 100 square feet of your treatment area to determine the number of gallons of water you will need:

0.01 gallon/sq ft × 100 sq ft = 1 gallon

Now you know that you need to dilute the 3 fluid ounces of pesticide into a minimum of 1 gallon of water in order to be able to apply it as a soil drench for 100 square feet of area.

The calculation may be different for a soil injection application depending on label instructions. If the label instructs you to apply the pesticide in sufficient water to allow you to place the same amount into each injection hole, follow the example below.

Continuing with the single-tree example described above, assume that you want to make four injections around the base of the tree. Knowing that you need enough solution to make 4 equal injections, you decide to mix one gallon of diluted material (there are 4 quarts in a gallon). This will allow you to apply 1 quart of this mixture to each injection hole. Since there are 128 fluid ounces of liquid in 1 gallon, you will need 125 fluid ounces of water mixed with the 3 fluid ounces of pesticide:

125 fl oz + 3 fl oz = 128 fl oz

Put 1 quart (32 fluid ounces) of this mixture into each of the four holes.

Calibrating Powered Equipment Used for Soil Injection

To determine the output of your application equipment, use a calibrated container. Fill the application tank with water, place the end of the injection probe into the calibrated container, and turn on the equipment. Measure the amount of output delivered in 1 minute. Repeat this step several times using the same pressure and the same injection probe. Find the mean of your measurements and use this as the delivery rate of your equipment in fluid ounces per minute.

In the example above, you determined that you need to apply a total of 128 fluid ounces or 4 quarts of diluted pesticide to the soil surrounding the base of a tree. You planned to make the application using four injection sites and put 1 quart of diluted pesticide into each hole.

Now you need to find out how long it will take your application equipment to deliver 1 quart of liquid. For this example, suppose you have determined that the output of your application equipment is 5 quarts per minute. To find out how long it takes to deliver 1 quart, divide 1 quart by 5 quarts per minute:

1 quart ÷ 5 quarts/minute = 0.20 minutes

Convert 0.20 minutes into seconds by multiplying 0.20 minutes by 60 seconds per minute.

0.20 minutes × 60 seconds/minute = 12 seconds

It will take 12 seconds to inject 1 quart of mixture into each hole.

Calibrating Hand-Held Injectors

Hand-held injectors have a range of output settings, usually between 1 and

5 ml (milliliters) or cc (cubic centimeters). One ml is virtually equivalent to one cc, and equal to approximately 0.033 fluid ounce. The label application instructions for the product you are applying will tell you what setting to use. A product label might, for example, direct you to set the injector to 5 ml in order to deliver ⅙ fluid ounce of solution per strike. This setting works out to 1 fluid ounce of solution per every 6 strikes. With this in mind, adjust the injector to the 5 ml or 5 cc setting and fill it with water. Place a block of wood or some other firm but not overly hard object in the bottom of a container. While resting the probe of the injector on the wooden block, make 12 strikes. Measure the water captured in the container. Do this two or more times and take a mean of the measurements.

Suppose, for example, that at the 5 cc setting 12 strikes release on average a total of 3 fluid ounces. The label application rate tells you that you need to inject a total of 30 fluid ounces of solution around the tree or shrub you are treating, and you have decided to make 5 injection holes. This means that you will need to inject 6 fluid ounces of the solution into each hole.

$$30 \text{ fl oz} \div 5 \text{ injection holes} = 6 \text{ fl oz/hole}$$

You have already determined that 12 strikes generally release 3 fluid ounces of solution. You also know that you need to inject 6 fluid ounces per hole to achieve your desired application rate. Therefore, you will need to apply 24 strikes per injection hole in order to dose each of the 5 holes with 6 fluid ounces of solution.

Calibrating Soil Drench Equipment

The calibration of soil drench equipment is very straightforward. Simply fill the delivery tank with water, turn the equipment on for a specific period of time, and measure the amount of output. Repeat these steps two more times to get a mean measurement for the delivery rate. Once you know the delivery rate and the amount of drench you wish to apply, you can figure out how long it will take to apply the desired amount of drench. For example, suppose you want to apply 6 gallons of soil drench per tree and your soil drench equipment delivers 2 gallons of drench solution per minute. Divide the desired drench amount by the applicator output rate to find out how long you must run the sprayer:

$$6 \text{ gal of drench} \div 2 \text{ gal/min} = 3 \text{ min application time}$$

Determining Doses for Tree Injection and Tree Implants

You can inject pesticides into a tree using capsules that contain a pre-measured amount of pesticide. When placing tree implants or injecting trunks with pre-measured capsules, you will need to determine how many capsules or implants to use, a number based on the diameter of the tree trunk's DBH. The label instructions for these products may specify required spacing intervals for the pesticide capsules. For example, a label may recommend that you space the pesticide capsules every 6 inches around the trunk of the tree. When you apply pesticides as slow-release implants, the capsules remain in the tree. Alternatively, when you inject pesticides from pre-measured capsules, you should remove the emptied capsules immediately after application is complete.

CHAPTER 9: REVIEW QUESTIONS

1. Increasing nozzle orifice size will produce

- ☐ a. smaller, uniform droplets
- ☐ b. a greater mixture of droplet sizes
- ☐ c. spray droplets that are more prone to drift
- ☐ d. larger droplets

2. If the pressure of your sprayer drops off, this could be an indication that the

- ☐ a. pump speed is too fast
- ☐ b. nozzles are too small
- ☐ c. pump is worn
- ☐ d. nozzles are mismatched

3. Frequent calibration of your application equipment will ensure that you

- ☐ a. are using the correct amount of pesticide for effective pest control
- ☐ b. can use the maximum amount of pesticide allowed by law
- ☐ c. will eradicate target pests
- ☐ d. can make effective pesticide applications during severe weather conditions

4. Why must you accurately measure the capacity of your spray tank?

- ☐ a. the tank manufacturer's capacity ratings may be inaccurate
- ☐ b. to be able legally to use the maximum pesticide product per acre as prescribed by the pesticide label
- ☐ c. to properly measure nozzle output
- ☐ d. to determine need for adjuvants

5. When making a pesticide application, an increase in the travel speed of the spraying equipment will

- ☐ a. increase the amount of pesticide applied per square foot
- ☐ b. decrease the amount of pesticide applied per square foot
- ☐ c. create larger droplets that are less prone to drift
- ☐ d. improve coverage of the pesticide being sprayed

6. Which of the following is not a good reason for calibrating pesticide application equipment?

- ☐ a. effective pest control
- ☐ b. faster equipment operation
- ☐ c. protecting human health, the environment, and treated surfaces
- ☐ d. complying with the law

7. For calibration of large turf application equipment, which four factors need to be measured?

- ☐ a. travel speed, sprayer pressure, nozzle height, and number of nozzles
- ☐ b. swath width, travel speed, sprayer pressure, and size of area to be treated
- ☐ c. swath width, travel speed, tank capacity, and sprayer output (flow rate)
- ☐ d. swath width, travel speed, size of area to be treated, and nozzle height

8. Your calibrated sprayer with a 30-gallon tank will cover 1.25 acres. You plan to apply an herbicide at a label rate of 1.5 pounds per acre. How many pounds of this herbicide will you put into the spray tank?

- ☐ a. 1.25
- ☐ b. 1.875
- ☐ c. 2.0
- ☐ d. 3.75

9. It takes your equipment 3 minutes to travel 264 feet. How fast, in miles per hour, is the equipment traveling?

- ☐ a. 1
- ☐ b. 2
- ☐ c. 3
- ☐ d. 4

10. By measuring the output of each nozzle on the spray boom, you discover that the sprayer output is 256 ounces in 30 seconds. What is the output of the sprayer in gallons per minute?

- ☐ a. 2
- ☐ b. 3
- ☐ c. 4
- ☐ d. 5

11. How many square feet can you treat if your sprayer holds 5 gallons and you've calibrated it to spray 0.02 gallons per square foot?

- ☐ a. 100
- ☐ b. 150
- ☐ c. 200
- ☐ d. 250

12. An advantage of using an RTU formulation is that

- ☐ a. it costs less
- ☐ b. larger amounts can be applied
- ☐ c. no mixing is required
- ☐ d. it is sold in recyclable containers

13. Using which type of herbicide application equipment would result in the least amount of drift?

☐ a. backpack sprayer
☐ b. small compressed-air sprayer
☐ c. wick applicator
☐ d. low-pressure sprayer

14. Which nozzle material would best resist abrasion?

☐ a. plastic
☐ b. brass
☐ c. aluminum
☐ d. stainless steel

10 Safe and Effective Applications in Landscape and Turf Settings

THIS CHAPTER PROVIDES ideas on how you can make safe and effective pesticide applications in outdoor landscape and turf settings. Chapter 11 will cover safe and effective applications in interiorscape settings.

People practice pest control in many different landscape and turf settings in California. These include home landscapes, golf courses, institutional and commercial grounds, greenbelts, cemeteries, parks, athletic fields, and school, college, and university grounds (Figure 10-1). They also include amusement parks, wildlife preserves, and preserved natural areas *(naturescapes)*. In these settings, turfgrass and landscape plants have a number of purposes, or "jobs," including

- making the area more attractive
- increasing the value of the property
- providing a high-quality, safe playing surface for sports
- controlling erosion
- making a playground safer for children's activities
- making a hot area cooler
- protecting soil and water

MAINTAINING HEALTHY TURF AND LANDSCAPES

Turfgrass and landscape plants do their jobs best when they are kept healthy. Pests are less likely to damage healthy turfgrass and landscape plants, and healthy plants have a better chance of surviving pest damage should it occur. One result of having healthier, more-resistant plants, then, is that you may end up needing to use fewer pesticides.

FIGURE 10-1.

Landscape pest control is carried out in many different settings, including parks, golf courses, and institutional grounds.

FIGURE 10-2.

To keep turfgrass and landscape plants healthy, you need to provide appropriate care in a timely manner. For turfgrass, this involves proper mowing, watering, aerating, dethatching, fertilizing, and topdressing.

FIGURE 10-3.

*For turfgrass maintenance and pest management, consult the UC IPM interactive **UC Guide to Healthy Lawns** at http://www.ipm.ucdavis.edu/TOOLS/TURF/ .*

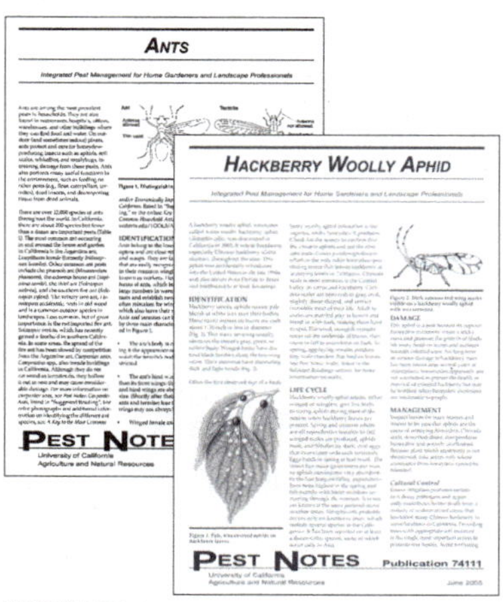

FIGURE 10-4.

*UC IPM **Pest Notes** cover a large number of pest management topics for landscape professionals and home gardeners. This resource is available at http://www.ipm.ucdavis.edu.*

To keep turfgrass and landscape plants healthy, you have to provide appropriate care in a timely manner. For turfgrass, this involves proper mowing, watering, aerating, dethatching, fertilizing, and topdressing (Figure 10-2). For landscape plants, this includes watering, mulching, fertilizing, and pruning. Although we mention some of these maintenance practices in this volume, we do not discuss them in detail. You can learn more about these practices from a number of other sources. For example, for turfgrass maintenance and pest management, consult UC IPM's interactive *UC Guide to Healthy Lawns* (Figure 10-3). In addition, UC IPM's *Pest Notes* cover a large number of pest management topics for landscape professionals and home gardeners (Figure 10-4). Both of these resources are available at http://www.ipm.ucdavis.edu.

SAFE AND EFFECTIVE TURF AND LANDSCAPE APPLICATIONS

Although keeping your turf and landscape plants healthy can help you keep pesticide use to a minimum, there may still be times when an application is necessary. As a pesticide applicator, you need to know how to make applications safely and effectively. To do this, you must

- read, understand, and follow all directions and warnings on the pesticide label
- read, understand, and follow all applicable pesticide use laws and regulations
- understand what will happen to the pesticides after you apply them (the *pesticides' fate*)
- know how to avoid pesticide drift
- be able to recognize common symptoms of phytotoxicity and know how to prevent it
- give advance notice to other workers and the public, and then keep people and their pets away from treatment areas during and after applications
- understand how to protect natural enemies and other beneficial organisms
- understand how to minimize leaching and runoff
- make sure that any building air intake vents near to the landscape or turf that you are treating are closed during application

Follow Label Requirements, Laws, and Regulations

Most turf or landscape pesticide applications are made in areas where there are people, pets, and buildings nearby. To handle pesticides safely in these settings, you must read and understand the entire pesticide label, including the label's use directions, before you begin to mix, load, or apply the chemical. You must also adopt safe pesticide handling habits and follow the appropriate standards of care for the pesticide applications you will make. You will find the standards of care on the pesticide label and in pesticide use laws and regulations. If you cannot read the label or laws and regulations, someone must read them to you and explain them.

The pesticide label is a legal, binding document. In some cases, California pesticide use laws and regulations may be more strict than the label, and in those cases you must follow whichever directions, requirements, and standards of care are most strict. The following list applies to the pesticide label *and* pesticide use laws and regulations. You must

- use pesticides only on plants, sites, or locations listed on the label (Ornamentals pose a unique problem because individual species are often not listed on labels. Unless a label gives a broad classification of *ornamentals, foliage plants,* or something similar, the pesticide cannot be used. When in doubt, check with the Department of Pesticide Regulation for a label interpretation.)

- wear all of the required personal protective equipment (PPE) (Figure 10-5)
- follow all directions about weather conditions and preventing drift or other types of off-site movement
- heed all warnings about phytotoxicity and other adverse effects
- understand the directions for mixing, loading, application, and disposal
- pay attention to label warnings about tank mixes and compatibility problems
- follow all directions and requirements that indicate which application equipment you should and should not use
- understand what to do in the case of a spill or other emergency
- keep people and pets out of the treatment area

See Chapter 4 in *The Safe and Effective Use of Pesticides, Second Edition,* for a detailed discussion of the parts of a pesticide label.

FIGURE 10-5.

When you mix, load, or apply pesticides in landscape settings, you must wear all of the label-prescribed personal protective equipment (PPE).

Pesticide Fate

You need to have some idea about the environmental fate of pesticides that you apply in turf and landscape environments. A pesticide's fate is what happens to it—where is goes, how it changes chemically—once you have applied it. Not all of the pesticide that you apply to a turfgrass or landscape planting gets onto the target leaves, roots, or other treated plant parts. There are several other things that can happen instead. The pesticide can

- be washed off of the plant and into the soil or turfgrass thatch by rain or irrigation water

- be washed off of the plant and carried away from the treated area into a storm drain or nearby creek, lake, pond, or other body of water (runoff)
- be chemically decomposed by the sun's rays (photodecomposition)
- travel down through soil layers and eventually contaminate ground-water (leaching)
- evaporate from the surfaces of treated plants or come out of (be desorbed from) the soil (In this case it would turn into a gas that would dissipate into the air *[volatilization]*. This gas can be carried a long way from the application site, and some of it may come back to earth in rainfall and thus pollute surface water.)
- remain on the plant as a residue and then be brushed off by people or animals when they enter treated areas (If this happens, pesticide residues can get on clothes, shoes, skin, animals' hair, or the bottoms of humans' or animals' feet.)
- become tightly attached to organic matter in the soil or in turfgrass thatch and stay there *(adsorption)*
- be chemically decomposed by microorganisms in the soil or in turfgrass thatch *(decomposition)*
- remain on grass clippings that then become part of commercial compost (In this case, sometimes the pesticide will not break down completely in the composting process, and desirable plants may be injured when they are treated with the contaminated compost [see Sidebar 30].)

Figure 10-6, for example, shows one possible fate for a pesticide after a turfgrass application.

Some things that happen to pesticides after you apply them are beyond your control. For instance, volatilization

SIDEBAR 30

Clopyralid and Commercial Compost

Most pesticides break down under normal composting conditions. If any pesticide remains, it is usually at such a low level of concentration that the compost is still safe to use. Clopyralid, however, is a particularly persistent herbicide and does not break down efficiently under normal composting conditions. Residues have been found in finished compost in California.

Clopyralid residues in compost that is used on plants such as beans, tomatoes, and eggplant can cause injury even at extremely low concentrations (1 part per billion). No damage from clopyralid-contaminated compost has been reported in California, but there have been reports of damage in other states.

The California Department of Pesticide Regulation has introduced restrictions on the use of clopyralid in California. It can no longer be used on residential lawns. In addition, there are restrictions on its use in parks, playing fields, and cemeteries.

For further information on clopyralid restrictions in California, visit the California DPR Web site at http://www.cdpr.ca.gov/ .

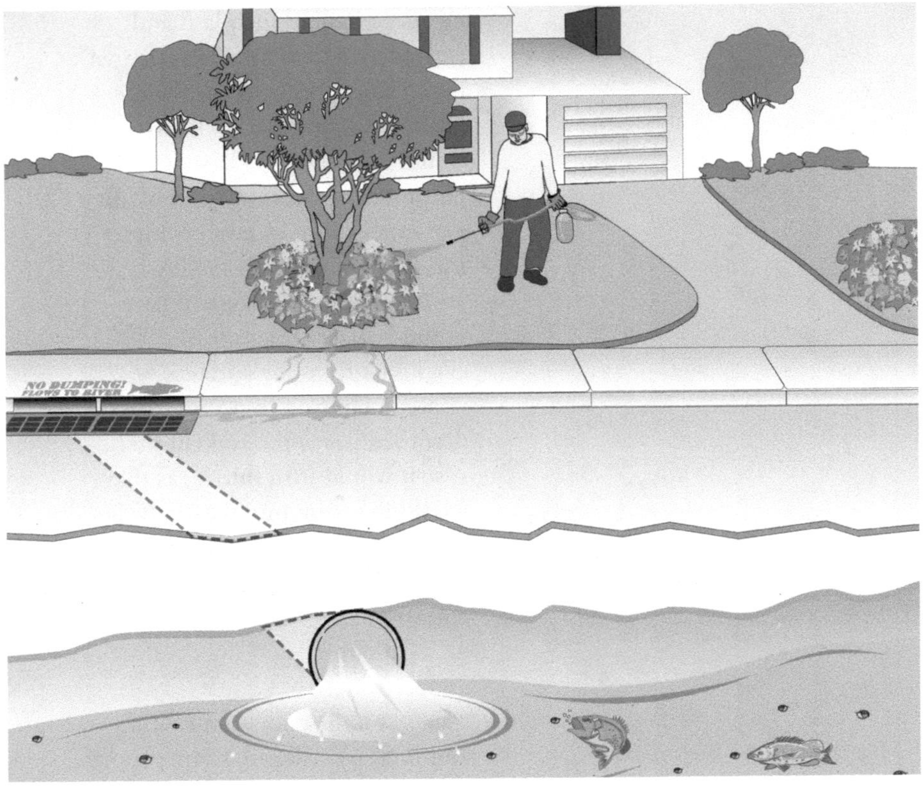

FIGURE 10-6.

Several things can happen to a pesticide after it is applied to landscape plantings. "Pesticide fate" refers to where the pesticide goes and what happens to it once you have applied it.

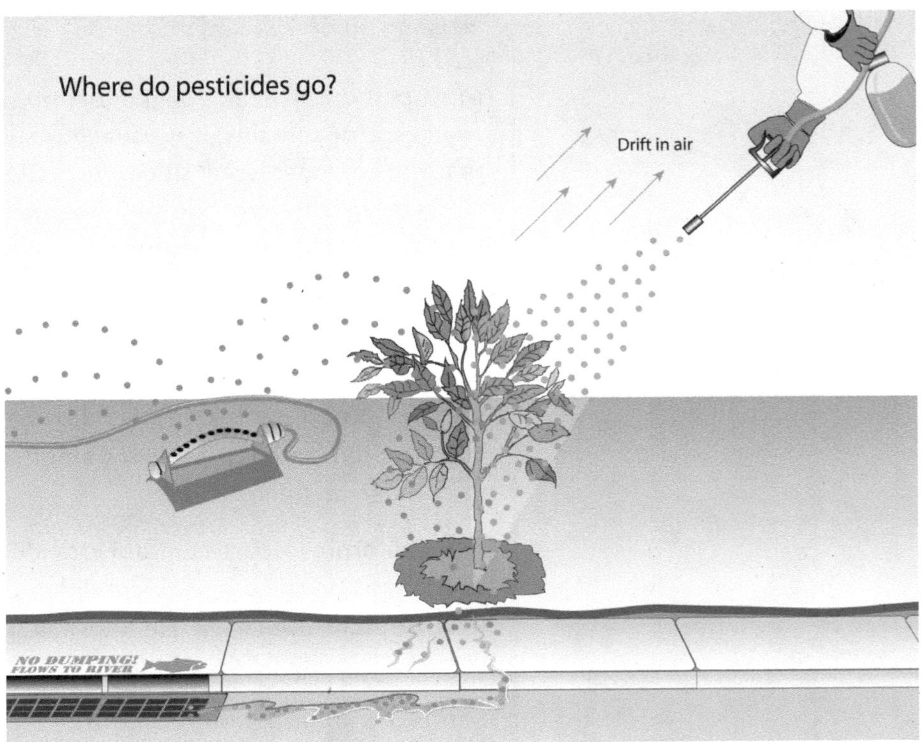

FIGURE 10-7.

Pesticides reach creeks and rivers through storm drains and household drains.

of a pesticide applied to plants, even after a safe and effective application, may continue for several days or weeks. Even in this sort of situation, however, you still have some ability and the responsibility to reduce the risks to people and the environment. For example, you can keep the leaves, thatch layer, and soil of a turfgrass sward healthy so that there is sufficient organic matter and beneficial microorganisms thrive. This helps in the processes of pesticide adsorption and decomposition. You must also make every effort to prevent any off-site movement of pesticides in order to keep them from

- running off into a storm drain or nearby body of water through the action of rainwater or irrigation water (Figure 10-7)
- leaching into groundwater
- spreading from the application site after they come into direct contact with people or animals that have entered the treated area too soon after treatment
- moving away from the application site through drift

Pesticide Drift

When you make a pesticide application, air movement might carry some of the pesticide away before it can land on the target. This is called *drift*. Be very careful to avoid drift during applications to turf or landscape areas. These settings often feature a variety of plants growing in close proximity to one another. A pesticide that is safe for one kind of plant might damage or kill a different kind of plant growing close by. In addition, people and animals are often near a landscape or turf application site.

Each time you make a pesticide application, carefully determine whether specific conditions at the application site will allow you to apply the pesticide safely. Prevent drift from affecting people, pets, automobiles, homes, furnishings, yards, or other property. Also, make sure that people remove or cover pet food dishes, bird's nests, bird feeders, and birdbaths in the area. Some landscape trees simply cannot be sprayed because the risk of harmful drift is too great. Their large size or their location near nontarget areas can make the potential for drift a very real concern. In these situations, pest managers sometimes use tree or soil injections since these do not pose the risk of drift, although they do create other problems.

Avoid Drift. To avoid causing pesticide drift, always

- check the weather forecast and make applications only under label-recommended weather conditions
- spray with nozzles at the proper height above the target
- use the correct sprayer pressure and nozzle orifice size
- leave an untreated buffer strip next to sensitive areas

Avoid Plant Damage

Pesticide phytotoxicity is the unintended chemical injury of plants and it is particularly detrimental when it occurs on valuable turf or landscape plants. Pesticides that may cause phytotoxicity include some herbicides as well as certain insecticides, such as oils.

Never use a pesticide that is not labeled for use on the plant species or groupings of plant types that you want to treat. Even when you are using them for plants or sites listed on the label, certain pesticides may cause injury if

- you use them on seedlings or flowers
- temperatures are too high during or immediately after the application
- plants are stressed for water or nutrients
- you use too much of the pesticide

FIGURE 10-8.

Most phytotoxic injury in landscape settings is caused by herbicide drift. Scorching and spotting of leaves are two symptoms of possible phytotoxicity.

- you make repeated applications
- you use the wrong formulation
- you mix them with incompatible materials

Herbicides are the most common cause of pesticide phytotoxicity (Figure 10-8). An herbicide may come into contact with nontarget turf or landscape plants in any or a number of ways, including

- drift
- spilling
- careless application
- spray application of residues from a previous herbicide application
- runoff

Damage to nontarget plants, including trees, can also occur when pesticides are taken up by the plants' roots. For example, roots can intercept an herbicide applied to control weeds in the area. In that case, the herbicide will move upward and reach the leaves, where symptoms of injury occur. A particular problem in settings such as golf courses occurs when tree roots intercept turf herbicides. Before you apply herbicides, make sure that you know the location of tree roots. Avoid applying herbicides near these roots.

Follow Label "Directions for Use" and Regulatory Requirements. You must carefully follow the "Directions for Use" on the pesticide label as well as any requirements that may be specified in regulations for certain types of pesticide, such as phenoxy herbicides. Under some conditions, selective herbicides can damage or kill desirable turfgrass or landscape plants even when you use them correctly. The label's "Directions for Use" relating to application rate, application timing, and any restrictions connected with drift are especially important (Figure 10-9). You must also follow the label's "Directions for Use" with regard to irrigation after an application. Irrigation may be required to activate the pesticide, but the wrong amount of water or poor irrigation timing might cause runoff, moving the pesticide to another area. This could cause damage to desirable turfgrass or landscape plants and may also cause environmental contamination.

FIGURE 10-9.

Read the pesticide label's "Directions for Use" relating to application rate, application timing, and any restrictions connected with drift before you make an application. This herbicide label warns you about the ways the herbicide can harm or kill nontarget plants if you apply it incorrectly.

FIGURE 10-10.

You must make sure that people and their pets are not in an area where you are going to apply a pesticide.

Recycling of irrigation water is now a common practice in some landscapes. If this is the case where you are managing pests, you will need to carefully track the use of herbicides and certain other pesticides so you can prevent these materials from getting into the recycled water.

Pesticide Use and Plant Maturity. Plants are most strongly affected by pesticides in the seedling stage. This is true for desirable plants as well as for weeds. Certain pesticides can seriously stunt or kill turfgrass seedlings and immature landscape plants. With few exceptions such as preemergent herbicides, only use pesticides on established turfgrass and landscape plants.

Herbicide Selectivity. Choose a selective herbicide whenever one is available, since phytotoxicity is a greater risk with nonselective herbicides. Be aware that when you use nonselective herbicides the material may stick to the soles of your shoes or to application equipment such as hoses or wheels (this may also occur with selective herbicides, but phytotoxicity is less of a concern than with nonselective pesticides). If this happens, you may find yourself tracking phytotoxic material into parts of a landscape planting where they could cause harm.

Spot-treatment is a good way to apply a nonselective herbicide to unwanted plants. Spot treatments are particularly useful against individual weeds or patches of weeds in small plots. They are also useful for applications around sprinkler heads, in sidewalk cracks, and along lawn edges. For spot spraying, you should apply just enough herbicide to wet the leaves of the target plant. If you spray too much, you can kill desirable turfgrass or other plants surrounding the sprayed spot. Spraying too much also puts unnecessary pesticide into the environment, and could increase risks of off-site movement.

Notify Other Workers and the Public

When you are about to make any kind of pesticide application in a turf or landscape setting, people in or near the treatment area need to know. You are legally responsible for notifying them. This includes notifying other workers, such as other pesticide handlers and mowers, and members of the public. They must be told to leave the area and take their pets with them, and you have to make sure that they do so before you perform the application (Figure 10-10). In addition, you must let them know how long to stay out. People and their pets must stay out of the treated area until the restricted-entry interval (specified on the pesticide label) is over.

If you perform tree injections, follow all label precautions to protect other workers and the public. Sometimes an injection device (such as a microinjector) must remain in the tree for a number of hours or even for a few days. During this time, you must ensure that people know to stay away and to keep their pets out of the treatment area. When the application is complete, immediately remove microinjectors and other application devices and dispose of them according to label directions.

There are special rules about notifying people verbally and about posting written warnings and other important information when you make a turf or landscape application. See Chapters 4 and 6 for information on California's specific legal notification requirements for turf and landscape settings.

Protect Beneficial Organisms

The environment includes many organisms that support the health of turfgrass and landscape plants. These beneficial organisms include

- *natural enemies* that attack turf or landscape pests
- *microorganisms* such as beneficial fungi and bacteria in soil
- earthworms and insects that burrow and create air spaces in the soil (These also help break down organic matter in soil and turfgrass thatch, improving soil structure.)
- birds, toads, honey bees, and various other organisms

Never use pesticides unless absolutely necessary: the overuse of pesticides can kill beneficial organisms, and this can weaken the overall health of turfgrass and landscape plants. Remember, when plant and soil health declines, turfgrass and landscape plants become much more vulnerable to pest attacks. If a pesticide application is necessary, spot-treat whenever possible.

Avoid Runoff and Leaching

Runoff from a turfgrass or landscape pesticide application may put nearby surface water at risk. Subsequent leaching of the pesticide may contaminate groundwater.

Runoff. When water is applied so quickly that the soil cannot absorb it all, the unabsorbed water becomes runoff. Both rain and irrigation can cause runoff. Some soils, such as clay soils or soils with high organic matter content, absorb water more slowly than do sandy or coarse soils. Less-absorbent soils also include soils that

- are already moist
- have a fine texture
- are compacted

Fine-textured soils have very small soil particles. Soil compaction occurs when the particles that make up the soil are pressed very tightly together. When this happens, it becomes hard for water to move through the soil because there are so few empty spaces for it to travel through. Runoff is more likely in these soils. Runoff is also more likely if the treated area is on a slope or if irrigation water is applied too fast for the soil to absorb.

Some types of pesticides are more apt to end up in runoff. Pesticides that dissolve in water *(water-soluble pesticides)* can wash off of plants and run off more easily. Pesticides that are more persistent (that stay longer in the environment) have a higher probability of ending up in runoff because there are more opportunities for them to be exposed to irrigation or rainfall. Runoff water transports liquid formulations more easily than granular formulations.

You can decrease runoff by

- making it easier for the soil to absorb water by amending the soil before you plant turfgrass or landscape plants or by topdressing amendments into existing turfgrass
- reducing soil compaction in turfgrass settings through regular aerations (Sidebar 31)
- regularly removing excessive thatch (more than ½ inch thick) from warm-season turfgrass so it will not block proper drainage (Sidebar 32)
- not overwatering turf or landscape plants, especially on slopes
- using the least water-soluble pesticides whenever possible
- directing sprinklers away from the treated parts of plants
- avoiding applications on or adjacent to bare soil surfaces
- providing a buffer strip or vegetative filter strip, for instance by not treating the bottoms of slopes, borders, or edges adjacent to hard surfaces

Leaching. In turfgrass plantings, certain turfgrass characteristics reduce pesticide leaching to a degree. Examples include

- high shoot density
- a thatch layer where some of the pesticide can get tightly bound to organic matter or get broken down by microorganisms
- a root zone that also hosts microor-

SIDEBAR 31

Turf Aeration

Aeration is a process that involves punching holes in the turfgrass and soil, either by machine or by hand. Aeration loosens compacted soil and reduces water runoff. It also helps to get more oxygen into the soil, which in turn encourages root growth and helps soil microorganisms to thrive.

Hollow-tine aerators (like the one shown above) create holes by removing cores or plugs of soil and thatch. The machine scatters the cores on the surface of the turfgrass as they are removed. You can use a rake to break these cores up and incorporate them into the turfgrass. For larger turfgrass areas, you can use a tractor to drag a piece of chain link fence across the turfgrass to break up and distribute the cores.

Solid-tine aerators create holes by pushing aside (displacing) the soil and thatch around the tine. This causes some compaction of the soil around the sides and bottom of each hole. You can aerate a very small turfgrass area using a hand-held spading fork, but again there will be some compaction. You can also use a hand-held sod-coring tool for small areas. This tool creates aeration holes by removing cores from the soil rather than by pushing the soil aside to create holes.

SIDEBAR 32

Thatch Removal

Thatch is the brown, tangled layer of dead and living roots, stems, ***stolons, rhizomes,*** and shoots that develops between the soil surface and the green part of warm-season turfgrass leaves (cool-season turf has no thatch). Whenever more than $1/2$ inch of thatch is present, you should remove some of it. An excessive thatch layer can make it very hard for water to get into the soil and reach the turfgrass roots. Excessive thatch can also hold water like a sponge. This creates conditions that are very hospitable to diseases and harmful insects.

Mechanical thatch removal is performed using vertical mowers. A vertical mower has revolving vertical blades that cut through the thatch and bring it to the surface. It can then be collected using a garden rake, a vacuum, or a sweeper. You can remove thatch for a small area by hand using a dethatching rake. You should dethatch when the turfgrass is growing vigorously since that is when it will be easiest for the turfgrass to recover.

ganisms that can break down some of the pesticide

In landscape plantings as well as turfgrass, the type of pesticide and the type of soil affect the extent of leaching. Highly water-soluble pesticides leach more easily. Sandy soils allow more leaching than clay soils. Soils with more organic matter help to break pesticides down before they can leach. The risk that pesticides will leach into groundwater is great if you apply a highly water-soluble pesticide to plantings on sandy soil. The risk is even greater if you overwater the plants or if they receive heavy rain.

With good management practices, you can help reduce pesticide leaching and the contamination of groundwater. Examples of management practices that can reduce leaching include

- selecting pesticides based on pesticide properties (such as how well the pesticide dissolves in water) and the treatment site's specific soil properties (such as soil texture and amount of organic matter the soil contains)
- timing applications to avoid periods of heavy rainfall or irrigation
- using extreme care when irrigating after applying highly water-soluble pesticides

CHAPTER 10: REVIEW QUESTIONS

1. It is important to understand what happens to pesticides after you apply them so you can

☐ a. use them more efficiently
☐ b. better protect the environment
☐ c. control pests more effectively
☐ d. make applications more cost effective

2. You must use pesticides only on the plants, sites, or locations listed on the label in order to

☐ a. comply with the law
☐ b. improve pesticide effectiveness
☐ c. reduce offsite movement
☐ d. protect nontarget organisms

3. The various things that happen to a pesticide after it is applied to a site are part of the pesticide

☐ a. history
☐ b. characteristics
☐ c. description
☐ d. fate

4. Rainfall after a pesticide application could result in

☐ a. pesticide residues moving offsite into creeks or rivers
☐ b. increased pesticidal activity
☐ c. more phytotoxicity
☐ d. increased residual activity

5. One of the impacts associated with pesticide drift in a landscaped area is

☐ a. the potential for damaging nontarget plants
☐ b. control of pests over a larger area than anticipated
☐ c. reduced runoff from irrigation
☐ d. more persistent residues on foliage

6. An effective way to apply nonselective herbicides to leaves of unwanted plants is by

☐ a. soil incorporation
☐ b. spot treatments
☐ c. high-pressure sprays
☐ d. granule application

7. When applying certain turf herbicides around trees, there is a potential risk that the herbicides may

☐ a. lose their effectiveness
☐ b. injure the turf
☐ c. not work properly
☐ d. injure the trees

8. Runoff from rainfall or irrigation is less common in

☐ a. sandy soils
☐ b. clay soils
☐ c. compacted soils
☐ d. soils that are high in organic matter

9. Which types of pesticides are more likely to run off a landscaped site?

☐ a. persistent pesticides
☐ b. oil-soluble pesticides
☐ c. volatile pesticides
☐ d. systemic pesticides

10. The potential for pesticide leaching is increased by

☐ a. drought
☐ b. high temperatures
☐ c. rainfall
☐ d. cool temperatures

11. Not treating the bottoms of slopes, borders, or edges adjacent to hard surfaces is one way you can reduce

☐ a. pesticide breakdown
☐ b. runoff
☐ c. leaching
☐ d. volatilization

12. A plant is at higher risk for phytotoxicity if it is

☐ a. growing rapidly
☐ b. well watered
☐ c. adequately fertilized
☐ d. water stressed

13. Photodecomposition of a pesticide is caused by

☐ a. wind
☐ b. heat
☐ c. rainfall
☐ d. sunlight

11 Safe and Effective Applications in Interiorscape Settings

PLANTS ARE WIDELY used as interior design features in shopping malls, hotels, resorts, restaurants, museums, government buildings, medical complexes, conference centers, banks, and many other indoor settings. Because of their close proximity to people and because the plants are usually located in enclosed areas, you must observe special precautions when applying pesticides to them (see Table 11-1). It is important to tailor each interiorscape pesticide application to match the unique situation of the application site.

Interiorscape pest management programs should focus on non-chemical control methods whenever possible (see Sidebar 33). Pest exclusion, mechanical methods such as physical removal of pests or infected plants, and biological controls offer an effective non-chemical way to manage interiorscape pests. Introduction of natural enemies of insect and mite pests has proven very successful, especially in atrium settings. Should a pesticide application be warranted, however, combine it with the effectiveness of

TABLE 11-1.

Concerns, Impacts, and Precautions to Take when Using Pesticides in Interiorscapes.

CONCERN	IMPACT	PRECAUTIONS
Damage to surfaces	Staining, etching, pitting; odors; liability.	Cover surfaces near application site; avoid drift.
Disruption of normal use of the area being treated	Inconvenience; mess; concern.	Communicate with facility managers and others so they will know what is happening, when it will happen, and how long it will take.
Exposure to people in the area	Potential illnesses or injuries; liability.	Apply pesticides when people are not present; communicate with people using the area; post or rope off the area being treated.
HVAC system	Contributes to moving pesticide residues around; residues may be picked up into air ducts.	If possible, shut down the system during application; close off nearby air registers; increase ventilation through doors or windows if possible.
Off-site movement: Drift	Pesticide spray droplets move off the application site and may injure people or objects; liability.	Control air movement in treatment area; avoid fine sprays; use alternative application methods such as soil drenches when possible.
Off-site movement: Other	Movement of pesticide residues in irrigation water or on plant parts; water contamination; liability.	Monitor irrigation and avoid runoff; bag and dispose of prunings, dropped leaves, and other unwanted materials in area where treatment was made.
Plant species differences	Possible injury of sensitive nontarget species	Read label for warnings; shield nontarget plants from application.
Residues	Public and workers may contact residues on foliage or soil; potential for illnesses or injuries; liability.	Minimize pesticide use; use short-residual materials; rely on non-chemical control methods.
Spills/accidents	Serious contamination of nontarget surfaces; liability.	Develop a plan for responding to spills or accidents; have cleanup material on site; keep people away; take responsibility for cleanup.
Vapors and odors	May cause illness or hysteria; liability.	Communicate with facility managers and the public; use low-odor formulations whenever possible; have proper ventilation during application and afterwards.

SIDEBAR 33

Non-Chemical Methods for Controlling Pests in Interiorscapes

- ☐ **Prevention.** Prevent pest problems by inspecting all plant material scheduled to go into an interiorscape setting, making sure they are pest and disease free. Keep plants that have pest problems or diseases away from healthy plants.
- ☐ **Exclusion.** Keep infested plants out of interiorscape plantings. Also, keep pests away from interiorscape plantings by creating barriers: for instance, screens over windows and other entrances, water in pans to protect potted plants from ant invasions, and sticky material painted onto lower limbs, stems, or trunks of plants or onto container edges to block crawling pests.
- ☐ **Physical Control.** Remove pests from smaller plants by washing or wiping foliage and stems.
- ☐ **Biological Control.** Introduce natural enemies of insect and mite pests and avoid using pesticides that would destroy these beneficial organisms.
- ☐ **Sanitation.** Prune off and destroy disease-infected parts of plants. Remove unhealthy plants. Remove items that attract pests or provide places for them to hide—this method is useful for discouraging rodents.
- ☐ **Trapping.** Use sticky traps for certain insects. Mechanical and sticky traps work well for control of small rodents.
- ☐ **Plant Health and Vigor.** Healthy plants can survive pest damage more easily. Maintain interiorscape plant health by providing proper amounts of water, fertilizer, and light and by keeping foliage dust-free with regular cleaning.

FIGURE 11-1.

Whenever possible, use pesticides that are less hazardous, such as soaps and oils, to reduce risks to people using the area.

non-chemical control methods to reduce the amount of pesticide you will require and to lessen risks to the public. Whenever possible, use less-hazardous pesticides such as soaps and oils (Figure 11-1).

HEALTHY INTERIORSCAPES

An interiorscape can be kept virtually free of pests, especially if you establish it with pest-free plants and soils in the first place. Pest problems can occur, however, when infested plants are introduced from greenhouses or other sources. Pests can also enter interiorscapes through heating, ventilation, and air conditioning (HVAC) systems, open doors or windows, or through various human activities (Figure 11-2).

The best strategy for creating and maintaining a healthy interiorscape is to exclude pests in the first place. Steps you can take to exclude pests include

- inspecting plants and growing media carefully for signs and symptoms of pests before you introduce them into the interiorscape (Figure 11-3)
- keeping unhealthy plants or those that show signs of pest damage out of the area until it is clear to you that no live pests are present
- maintaining interiorscape plants' health by providing them with proper amounts of water, fertilizer, and light and by keeping their foliage dust-free with regular cleaning (Avoid cleaning methods that can spread pests to other plants, such as feather dusters [Figure 11-4])

Healthy interiorscape plants are more likely to survive pest attacks, but it is important that you monitor interiorscape plants for any sign of pests at every service visit. If you do find any sign of pests or pest damage,

- clean affected plants with disposable wipes (e.g., facial tissue)
- prune out infested portions of the affected plant
- remove infested portions or entire plant from the area before pests can spread to any healthy plants

FIGURE 11-2.

Pests can enter interiorscapes through heating, ventilation, and air conditioning (HVAC) systems, open doors or windows, or as a result of human activities.

FIGURE 11-3.

Carefully inspect plants and growing media for the presence or symptoms of pests before you introduce new plants into an interiorscape.

FIGURE 11-4.

The use of feather dusters to clean plant leaves can move pests to other plants. Disposable wipes are a far better choice.

OFF-SITE APPLICATIONS

Some of the problems associated with interiorscape applications include disruption and mess in the area, possible damage to carpets, draperies, and furniture, and potential danger to workers and the public. When you must use pesticides to control pests on interiorscape plants, the easiest way to avoid these problems is to remove the plants and treat them in a separate, controlled area. This may be an outdoor location, a greenhouse, or some other area. Apply only pesticides that are labeled for use on the plant or plants you are treating and labeled for indoor, interiorscape, or atrium use. If you are holding the plants in a greenhouse, make sure to use only pesticides that are labeled for use in a greenhouse.

Wherever you make the application, keep people away from the treatment area. Follow all label directions about restricted-entry intervals and personal protective equipment. Leave treated plants in the controlled area until it is safe to reintroduce them into the interiorscape. Consult the pesticide label for the length of time the plants should remain in the controlled area.

IMPROVING THE SAFETY OF ON-SITE APPLICATIONS

Sometimes it is not possible to remove plants from their interiorscape setting before treatment. For instance, very large potted plants and those in built-in planting beds must usually be treated where they are. When this is the case, it is best to conduct the application during non-business hours when members of the public and building staff are not around. Notify building owners, facility managers, or those in charge of building access about when you plan to make the pesticide application. You must also communicate clearly with them about any restricted-entry intervals (REI) or other information as required in regulations. Provide your clients with a Material Safety Data Sheet (MSDS) for the product being used. Use only pesticides that are labeled for use indoors on interiorscapes. Besides being illegal, the use of pesticides that are not labeled for interiorscapes is hazardous to the applicator and to the public.

Specific issues of safety, concern, and effectiveness arise when you are making pesticide applications to indoor

FIGURE 11-5.

Sometimes it is necessary to confine sprays to just the infested plant species and prevent drift onto other nearby plants. This is an especially important consideration when several species of plants are growing in the same container.

plantings. For example, members of the public, including children, are likely to frequent many of these areas, so you must protect them from exposure to vapors or residues during and after application. Other species of indoor plants may be closely spaced with the plant that requires treatment, so you may have to develop ways to confine your spray application to the infested plant species and prevent drift or other off-target movement of the pesticide (Figure 11-5). Pesticide spills, splashes, and drift can cause expensive damage to benches, walls, floors, carpets, paneling, and other sensitive surfaces near the planting, so you need to protect these areas. In addition, pesticide spray or vapors may be drawn into the building's HVAC system unless you arrange to have the system shut down or to block off vents near the application area.

To perform a pesticide application safely and effectively in any interiorscape setting, you must

- read and follow all directions and warnings on pesticide labels and in regulations
- understand the fate of the pesticide in interiorscapes
- know how to protect target and nontarget plants from phytotoxic damage
- know how to protect other workers and the public from pesticide exposure
- know how to protect nontarget surfaces

Following Label Directions and Regulation Requirements

Before you make any type of interiorscape application, read the pesticide label and review the regulations. During mixing, loading, and application, follow all label directions and warnings. Allow for enough time after application to comply with restricted-entry requirements. Thoroughly incorporate any granular formulations into the soil and then water the soil. Conceal spikes or baits so they will not attract the attention of children or pets.

Regardless of the formulation or application method you use, make sure to ventilate the treatment area thoroughly after any pesticide application to remove fumes or vapors before you allow people to enter. Do this by opening doors and windows, if possible. Use fans to move the air outside, but at the same time avoid getting any spray or vapor into the building HVAC system.

Mixing and Loading. If you are not using a ready-to-use (RTU) formulation, follow the label rates to accurately mix only enough pesticide for the current application. Depending on the size of your spray tank, some pesticide labels could call for amounts as small as 1/8 of a teaspoon or less mixed with water in the tank to cover the area or plants you are spraying. When working with such small quantities, it is easy to double or even triple the rate if you do not measure materials accurately. Table 11-2 is a conversion chart that you can use when preparing small quantities of pesticide mixture.

Perform pesticide-mixing activities outside, away from the building where the plants to be treated are located. This precaution eliminates the chance that you will spill concentrated materials in the building. See Sidebar 34 for mixing procedures.

Pesticide Fate in Interiorscape Applications

For interiorscape pesticide applications, issues of pesticide fate chiefly involve drift, residues, and off-site movement.

Drift. Pesticide drift during the application of foliar sprays can occur when air currents from heating and air conditioning vents or open windows and doors move droplets away from the target plants.

Residues. Residues are the pesticide materials that remain on foliage, on the surfaces of the growing medium, pots, and planters, and on other surfaces that might be sprayed during application or might be affected by drift.

Off-site Movement. Traces of pesticide may be carried into drains with irrigation water runoff or they may be carried off when plant parts, planting media, or planting containers are moved.

A limited number of application methods are used on interior plantings, and different issues of pesticide fate are associated with each. Interiorscape applications either treat the aboveground parts of plants or the growing medium. Applications used on leaves and stems include foliar sprays and dips. Typical soil treatments include drenches, granular applications, and slow-release spikes. Incorporation of the chemicals into the growing medium (e.g., incorporation of granules into the soil) reduces the risk of human exposure to residues unless these leach out into the liquid that drains from containers or planter boxes after watering. If the soil may contain pesticide residues, people involved in maintaining the plants must be made aware or they may

TABLE 11-2.

Spray Quantity Conversion Chart.

Use this chart to prepare small quantities of spray mixture when label directions instruct you to mix a measured amount of pesticide product in a certain volume of water. This chart will NOT work when directions require that you apply a measured amount of pesticide product to a certain area.

On this chart, find the recommended amount of chemical and the volume of water listed on your pesticide label. Then go left or right in that row to find how much pesticide product you will need for the volume of spray you wish to prepare. For example, if the label recommends 2 gallons of pesticide product per 100 gallons of water and you only want 2 gallons of spray, you will need 10 tablespoons (T) of pesticide product.

VOLUME OF PESTICIDE FORMULATION TO DILUTE IN *X* GALLONS OF WATER*†						
X = 100 gal	X = 50 gal	X = 25 gal	X = 15 gal	X = 5 gal	X = 2 gal	X = 1 gal
Liquid Formulations						
2 gal	1 gal	2 qt	2.4 (2¼) pt	1.6 (1½) c	10.24 (10) T	5.12 (5) T
1 gal	2 qt	1 qt	2.4 (2½‡) c	0.8 (¾) c	5.12 (5) T	2.56 (2½) T
2 qt	1 qt	1 pt	1.2 (1¼‡) c	6.4 (6¼) T	2.56 (2½) T	3.84 (3¾) t
1 qt	1 pt	1 c	9.6 (9½) T	3.2 (3) T	3.84 (3¾) t	1.92 (2‡) t
1 pt	½ pt	½ c	4.8 (4¾) T	4.8 (4¾) t	1.92 (2‡) t	0.96 (1‡) t
¾ pt	¾ c	6 T	3.6 (3½) T	3.6 (3½) t	1.44 (1½‡) t	0.72 (¾‡) t
½ pt	¼ c	4 T	7.2 (7) t	2.4 (2½‡) t	0.96 (1‡) t	0.48 (½‡) t
Dry Formulations						
6 lb	3 lb	1½ lb	0.9 (1‡) lb	4.8 (5‡) oz	1.92 (2‡) oz	0.96 (1‡) oz
5 lb	2½ lb	1¼ lb	12 oz	4 oz	1.6 (1½) oz	0.8 (¾) oz
4 lb	2 lb	1 lb	9.6 (9½) oz	3.2 (3¼) oz	1.28 (1¼) oz	0.64 (⅔‡) oz
3 lb	1½ lb	12 oz	7.2 (7¼‡) oz	2.4 (2½‡) oz	0.96 (1‡) oz	0.48 (½‡) oz
2 lb	1 lb	8 oz	4.8 (5‡) oz	1.6 (1½) oz	0.64 (⅔‡) oz	0.48 (½‡) oz
1 lb	8 oz	4 oz	2.4 (2½‡) oz	0.8 (¾) oz	0.32 (⅓‡) oz	0.16 (⅙‡) oz
12 oz	6 oz	3 oz	1.8 (1¾) oz	0.6 (⅔‡) oz	0.24 (¼‡) oz	0.12 (⅛‡) oz
⅔ lb (10.66 oz)	5.33 (5⅓) oz	2.67 (2⅔) oz	1.6 (1⅗) oz	0.53 (½) oz	0.21 (⅕) oz	0.11 (⅒) oz
8 oz	4 oz	2 oz	1.2 (1¼‡) oz	0.4 (½‡) oz	0.16 (⅙‡) oz	0.08 (1/12‡) oz

* Liquid and dry equivalents: gal = gallon (128 fl oz); qt = quart (32 fl oz); pt = pint (16 fl oz); c = cup (8 fl oz); T = tablespoon (½ fl oz); t = teaspoon (0.167 fl oz); fl oz = fluid ounce; lb = pound (16 oz); oz = ounce

†Values shown in **bold** are exact amounts. The numbers in parentheses are approximations and may be slightly more or less than the exact amounts.

‡Using these double-daggered approximations will result in a more-concentrated mixture. DO NOT use these approximations unless the label instructions allow for a more-concentrated mixture.

SIDEBAR 34

Procedures for Mixing Pesticides

Pesticides must be properly mixed to ensure that the correct amount of pesticide is thoroughly incorporated into a measured amount of water or other solvent. Techniques for mixing pesticides are the same for large and small volumes. In interiorscape situations, though, very small amounts of active ingredient are common so be sure to have measuring equipment that is suitable for these small quantities.

Before beginning, read the mixing directions on the label of the pesticide you will be using. If adjuvants are needed, these are usually added before the pesticide products unless label instructions give a different order.

General rules for mixing pesticides include

- ☐ Determine what personal protective equipment (PPE) is required for mixing the pesticide by checking the pesticide label *and* the California regulations (see Chapter 6).
- ☐ Before you add a pesticide to the spray tank, check for leaks in the tank and hoses. Make sure the equipment is clean and operating properly.
- ☐ Use only clean water in the spray tank. Make sure that the pH of the water is within a range suitable for the pesticide(s) being used (see Chapter 3). You may need to add buffers or acidifiers to adjust the pH.
- ☐ Measure pesticides carefully, accurately, and safely to be certain that the correct amount of pesticide product is put into the spray tank (see the Calibration section in Chapter 9).
- ☐ When mixing more than one type of pesticide formulation into the spray tank, add the materials in the following order:
    1. Wettable powders
    2. Flowables and dry flowables
    3. Water-soluble concentrates
    4. Emulsifiable concentrates

unwarily touch the soil to check soil moisture, to plant new plants, or to care for existing ones. They must wear chemical-resistant gloves during any contact with treated soil (Figure 11-6).

When applied properly, granules are among the lowest-risk forms of pesticides for the applicator, the public, and the plants. The reduced amount of dust and lack of pesticide spray or mist lower the risk of inhalation exposure during application. All the same, you should avoid skin contact with the granules or with treated soil during application. Some granular pesticides are very toxic and easily absorbed through the skin. Wear chemical-resistant gloves during a granular pesticide application and when you work with treated plants or soil after an application.

Application of granules in interiorscape situations usually involves the even sprinkling of a measured quantity of granules from a small container onto the surface of the soil around the target plant. Once the proper amount has been applied, you incorporate the granules into the soil using a cultivation tool. Label recommendations may suggest that you irrigate the soil after application to start the granule breakdown process and so make the pesticide material available for uptake by plant roots.

Once you properly incorporate the granules into the growing medium, there is little danger of public exposure to the chemicals unless they leach, after watering, into saucers at the base of containerized plants or out of container or planter box drains.

Preventing Drift. Foliar sprays can drift on air currents from heating and air conditioning vents or from windows and doors and can remain as residues on nontarget plants, potting media, and nontarget surfaces. Ventilation is important when you are making an application, but you can avoid most

FIGURE 11-6.

Because the soil in a planted area may contain pesticide residues, people involved in caring for the plants must wear chemical-resistant gloves if their activities bring them in contact with treated soil.

FIGURE 11-7.

To avoid the possibility of drift during a pesticide application, make sure to use equipment that produces large spray droplets.

drift by adjusting the air vents or shielding the application area from air currents. Use low-pressure and large nozzle sizes to keep the spray droplets large and so less prone to drift (Figure 11-7).

Electrostatic sprayers used in large areas such as atriums eliminate runoff and greatly reduce the volume of pesticide applied. An electrostatic sprayer applies pesticides in the form of small, electrically charged droplets. Droplets average about 50 microns in diameter and receive a negative electrostatic charge as they leave the sprayer. Because plant material has a positive electrostatic charge, plant surfaces attract the spray droplets. The negatively charged spray droplets repel one another, so they do not clump together into larger droplets. These sprayers appear more effective if the spray droplets only travel a short distance from the sprayer to the target. As the distance increases, the electrostatic effect diminishes. The electrically charged droplets nearly eliminate runoff and improve coverage, greatly reducing the volume of pesticide required.

Avoiding Phytotoxicity

Phytotoxicity may be a problem with particular pesticides on certain sensitive plants. Sometimes the pesticide's active ingredient is not what causes the phytotoxicity. It may instead be the plant's reaction to solvents in the formulation or to impurities (such as salts) in the water that was mixed with the pesticide. Excessive application rates, inadequate mixing, and improper pesticide dilution may also cause plant damage. Some plants may be especially sensitive to injury from pesticide sprays or soil applications of certain pesticide materials. Check the pesticide product label for any precautions regarding sensitive plant species.

Environmental conditions such as temperature and humidity at the time of application can influence phytotoxicity. For most products, the risk of phytotoxicity increases as the temperature rises. Always apply pesticides when interiorscape conditions favor the rapid drying of sprays and when temperatures are at suitable levels for pesticide efficacy. Plants that are stressed due to water or nutrient deficits may be more susceptible to injury.

FIGURE 11-8.

When several species of plants are growing in one container, read and understand the label's phytotoxicity warnings before making drench applications.

When making foliage applications, you may find it necessary to shield sensitive plants while you spray other plants. In permanent planting areas, planter boxes, and large pots that hold several different types of plants, a pesticide applied as a soil drench to target plants can easily move through the planting medium to the roots of nontarget plants and may cause phytotoxic damage. Read and understand the label's phytotoxicity warnings before you make any drench applications to planters that contain more than one type of plant (Figure 11-8).

Notifying Other Workers and the Public

The best approach to protecting people is to make your pesticide applications during non-business hours. If other workers or the public are likely to enter the building before sprays have dried, you must post the treated area so that people will not enter it or allow their children or pets to enter. Follow all label directions and federal and state laws and regulations regarding notification and posting of interiorscape application sites. See Chapter 6 for posting and notification requirements. Provide building managers with a Material Safety Data Sheet for the product you are using (Figure 11-9).

Protecting Nontarget Surfaces

Some pesticides and adjuvants can cause severe, permanent damage to surfaces like linoleum, wood paneling, paint, and carpeting. You are responsible for preventing this damage, so protect all nontarget surfaces in interiorscapes with disposable coverings before you make an application.

Dealing with Pesticide Spills

Always take any spill of a pesticide material seriously. Before working with any pesticide, become familiar with the emergency procedures listed on the label. Know what personal protective equipment is needed for cleaning up spills. Thoroughly clean up and decontaminate anything that the spilled material contacted. Keep people away from the spill area until the cleanup is finished. For information on cleaning up spills, refer to the Material Safety

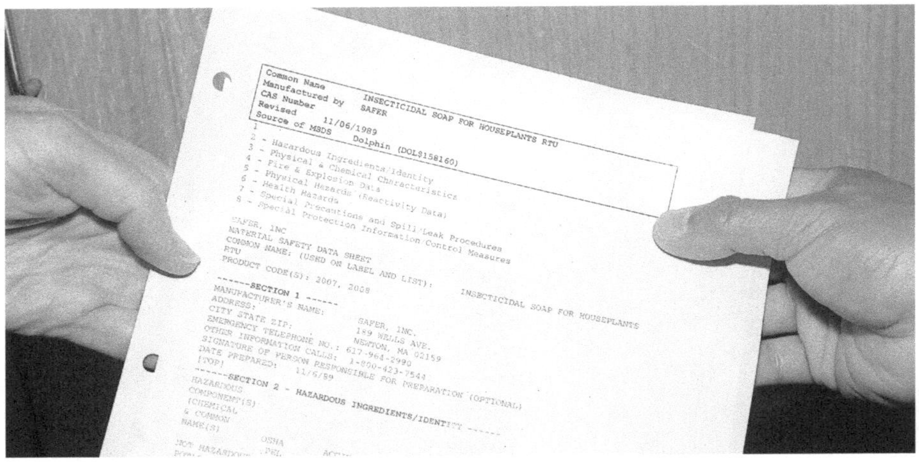

FIGURE 11-9.

Provide building managers with a Material Safety Data Sheet (MSDS) for the pesticide product being used.

Data Sheet for the pesticide that was spilled. The following are steps to take when a spill occurs:

- **Clear the Area.** Keep people away from the contaminated area. Provide first aid to anyone who has been injured or contaminated. Send for medical help if necessary.
- **Wear Personal Protective Equipment (PPE).** Before you begin any cleanup, put on the PPE listed on the spilled pesticide's label for handling the concentrated material. Check the pesticide label for additional precautions.
- **Contain the Spill.** Use soil, sand, sawdust, or absorbent clay to form a containment "dam" around liquid spills. Common cat litter is a good absorbent material for pesticide cleanup.
- **Clean up the Pesticide.** Proceed to clean up the spill. Brush the dam of absorbent material toward the center of a liquid spill. Add additional absorbent material if necessary. If the spill is on soil, shovel out contaminated soil for disposal. Remove any soil that becomes wet from a liquid spill. Large spills may require professional help to test for soil contamination. Place the absorbent material or spilled dry product and any contaminated soil in a sealable container and close it. Containers for contaminated materials must be suitable for transportation and must be labeled with the pesticide's name and its signal word.
- **Clean Nonporous Surfaces and Safety Equipment.** If the spill occurred on a cleanable surface such as concrete or asphalt, use a broom to scrub the contaminated surface with a strong detergent solution. Again, collect this wash water with absorbent material and place it in the sealable container. Be careful not to spill any contaminated material outside of the sealable container. Clean or dispose of any equipment such as brooms, shovels, and dustpans that you have used for cleanup. Close and seal the waste container. When you finish with the cleanup, clean the outside of your own PPE.
- **Dispose of the Material.** Local regulations on the disposal of hazardous materials may vary. Check with the local county agricultural commissioner or Water Quality Control Board for instructions on how to dispose of the sealed container and its contents.

CHAPTER 11: REVIEW QUESTIONS

1. Exclusionary tactics for managing pests in interiorscapes include

☐ a. use of pest-free plants and soils
☐ b. use of proper irrigation and fertilizer practices
☐ c. use of biological controls
☐ d. all of the above

2. Selective pruning can be an effective pest management method because pruning

☐ a. stimulates new plant growth
☐ b. can remove diseased parts of plants
☐ c. makes plants less attractive to pests
☐ d. changes the plant's resistance to pests

3. The main advantage to moving interiorscape plants offsite to apply a pesticide is that

☐ a. it is less expensive
☐ b. the applicator can use less-restrictive personal protective equipment
☐ c. you can use preferred agricultural-use pesticides
☐ d. it is less hazardous for the public

4. The best time to make onsite pesticide applications in a commercial facility would be

☐ a. early morning
☐ b. midday
☐ c. late afternoon
☐ d. during non-business hours

5. What information should you provide to a facility manager about a pesticide product before you make an application?

☐ a. the pesticide label and restricted-entry intervals
☐ b. the Material Safety Data Sheet and restricted-entry intervals
☐ c. a pesticide application check list
☐ d. the Pesticide Safety Information Series

6. It is desirable to adjust air vents or shield the application area from air currents to

☐ a. allow the spray to dry properly
☐ b. prevent the spray from drying too quickly
☐ c. reduce the chances of drift
☐ d. reduce the chances of an explosion

7. A restricted-entry interval (REI) is the period

☐ a. during an application when the public is not allowed to enter the treated area
☐ b. after an application when the public is not allowed to enter the treated area
☐ c. after an application when the public is allowed to enter the treated area on a restricted basis
☐ d. during and after an application when the public is allowed to enter the treated area on a restricted basis

8. Ventilating a treated area after a pesticide application is a good idea because it

☐ a. speeds up the drying of the pesticide
☐ b. activates the pesticide
☐ c. improves pesticide coverage
☐ d. removes odors and vapors

9. The major issues of pesticide fate in interiorscape situations involve

☐ a. drift and residues
☐ b. odors and half-life
☐ c. efficacy and persistence
☐ d. phytotoxicity and nontargets

10. When properly applied, which of the following pesticide formulation has the lowest risk?

☐ a. liquid sprays
☐ b. dusts
☐ c. aerosols
☐ d. granules

11. Hazards associated with granule applications or soil drenches include

☐ a. residues on the foliage
☐ b. vapors
☐ c. drift
☐ d. residues in the planting medium

12. For pesticide phytotoxicity information, you should check the

☐ a. pesticide label
☐ b. Material Safety Data Sheet
☐ c. UC IPM Pest Notes
☐ d. DPR Pesticide Safety Information Series

13. In a planting with mixed species of plants, you can minimize phytotoxicity by

☐ a. making multiple applications at lower doses
☐ b. increasing the application rate but not the application frequency
☐ c. using adjuvants to speed uptake
☐ d. shielding sensitive plants

14. The risk of phytotoxicity often increases as

☐ a. spray pressure increases
☐ b. spray mixtures become more dilute
☐ c. temperature rises
☐ d. day length increases

15. To protect surfaces surrounding the plants you will be spraying,

- ☐ a. use only water-soluble formulations
- ☐ b. avoid emulsifiable concentrates
- ☐ c. cover surfaces with disposable coverings
- ☐ d. increase the dilution rate

16. Pesticide-specific spill cleanup information is contained in the

- ☐ a. Material Safety Data Sheet
- ☐ b. pesticide product guide
- ☐ c. Pesticide Safety Information Series
- ☐ d. pesticide label

ANSWER SHEET FOR REVIEW QUESTIONS

Chapter 1
1. d
2. b
3. d
4. c
5. d
6. a
7. a
8. b
9. b
10. d
11. c
12. a
13. c
14. d
15. c
16. d
17. b
18. c

Chapter 2
1. a
2. c
3. a
4. c
5. a
6. d
7. a
8. a
9. b
10. c
11. d

Chapter 3
1. b
2. c
3. d
4. d
5. c
6. b
7. b
8. b
9. b
10. c
11. d
12. a
13. d
14. c

Chapter 4
1. d
2. c
3. c
4. a
5. a
6. c
7. d
8. b
9. b
10. b
11. b
12. a
13. d

Chapter 5
1. b
2. c
3. a
4. c
5. c
6. c
7. b
8. a
9. c
10. b
11. a
12. b
13. d
14. b
15. b

Chapter 6
1. d
2. c
3. a
4. d
5. c
6. b
7. a
8. b
9. d
10. d
11. a
12. c
13. b
14. c
15. b
16. b
17. a

Chapter 7
1. b
2. d
3. a
4. d
5. c
6. b
7. c
8. a
9. d
10. b
11. d
12. b
13. c

Chapter 8
1. a
2. a
3. c
4. d
5. d
6. a
7. d
8. b

Chapter 9
1. d
2. c
3. a
4. a
5. b
6. b
7. c
8. b
9. a
10. c
11. d
12. c
13. c
14. d

Chapter 10
1. b
2. a
3. d
4. a
5. a
6. b
7. d
8. a
9. a
10. c
11. b
12. d
13. d

Chapter 11
1. a
2. b
3. d
4. d
5. b
6. c
7. b
8. d
9. a
10. d
11. d
12. a
13. d
14. c
15. c
16. a

Glossary

abiotic. nonliving factors, such as wind, water, temperature, or soil type or texture.
absorb. to soak up or take in a liquid or powder.
acaricide. a pesticide used to control mites.
accidental misapplication. an unintentional, incorrect application of a pesticide.
accumulate. to increase in quantity within an area, such as in the soil or tissues of a plant or animal.
acidic. pertaining to a solution or substance that has a pH lower than 7.0.
acidifier. an adjuvant used to lower the pH (or *acidify*) the water being mixed with a pesticide. Pesticides often break down more slowly if the spray water is slightly acid. Also known as *acidulators.*
acidulator. see *acidifier.*
Acord Certificate of Liability Insurance. one of the form options required by the Department of Pesticide Regulation to verify that landscape maintenance pest control businesses and maintenance gardener business have liability insurance covering their pesticide use operations. The insurance provider issues the Acord Certificate of Liability Insurance.
action threshold. in pest management, the level of pest damage or pest infestation that warrants some type of control action.
activator. an adjuvant that increases the activity of a pesticide by reducing surface tension or speeding up penetration through insect or plant cuticle.
active ingredient (a. i.). the material in the pesticide formulation that actually destroys the target pest or performs the application's desired function.
acute effect. an illness that becomes apparent soon after an organism is exposed to a pesticide.
acute onset. pertaining to symptoms of pesticide-related injury that appear soon after the exposure incident.
adjuvant. a material added to a pesticide mixture to improve or alter the deposition, toxic effects, mixing ability, persistence, or other qualities of the active ingredient.
adsorb. to take up and hold on the surface of a plant or other organism.
aerating. a process used to reduce soil compaction in turf. Two processes are used—one uses metal prongs to create holes in the turf and the other uses hollow tubes to remove plugs of soil from the turf. Also known as *aeration.*
aerosol. very fine liquid droplets or dust particles, often emitted from a pressurized can or other aerosol-generating device.
aestivation. dormancy during summer or periods of high temperature or during a dry season. Sometimes spelled *"estivation."*
agitator. a mechanical or hydraulic device that stirs the liquid in a spray tank to prevent the mixture from separating or settling.
agricultural commissioner. the official in each county in California who has the responsibility for enforcing state and federal pesticide regulations and issuing permits for restricted-use pesticides. County agricultural commissioners and their staff frequently inspect pesticide applications and application sites. All agricultural uses of pesticides

and some non-agricultural uses must be reported monthly to the county agricultural commissioner.

agricultural use. a classification of certain pesticides that limits their use to production agriculture settings, including landscape and turf.

air gap. a space between the filling hose and the liquid in the pesticide tank that prevents the backflow of pesticide liquids into the water source. The air gap should be a minimum of 2 times the diameter of the filling hose or pipe.

algae. aquatic, nonvascular plants (singular: *alga*).

alkaline. pertaining to a solution or substance that has a pH greater than 7.0.

amending. adding materials, such as nutrients or organic material, to the soil in order to make up for deficiencies.

amphibian. a cold-blooded organism such as a frog, toad, or salamander.

anemometer. an instrument used for measuring wind speed.

animal kingdom. one of the major groups of living organisms, another being the plant kingdom.

annual. a type of plant that passes through its entire life cycle in one year or less.

annual weeds. weed species that have an annual life cycle.

anticoagulant. a type of rodenticide that causes death by preventing normal blood clotting.

application frequency restriction. a limitation on the number of times a particular pesticide may be applied to the same site during a given growing season or other period.

application pattern. the course the applicator follows through the area being treated with a pesticide.

application rate. the amount of pesticide that is applied to a known area, such as 1,000 square feet.

application swath. see *swath* and *swath width*.

aquatic. pertaining to something that grows or lives in water or is frequently found in water.

aquifer. an underground formation of sand, gravel, or porous rock that contains water. The place where groundwater is found.

arthropod. an animal having jointed appendages and an external skeleton, such as an insect, a spider, a mite, a crab, or a centipede.

artificial respiration. see *rescue breathing*.

attractant. a substance that attracts a particular species of animal. When manufactured to attract pests to traps or poisoned bait, attractants are considered pesticides.

augmentation. the process of building up a population of natural enemies in an area by bringing in additional eggs, larvae, or adults of that species.

avicide. a pesticide used to control pest birds.

back siphoning. the process that permits pesticide-contaminated water to be sucked from a spray tank back into a well or other water source. To prevent back siphoning, you can provide an air gap or check valve in the pipe or hose used to fill a spray tank.

backflow. see *back siphoning*.

backpack sprayer. a small, portable sprayer carried on the back of the person making the pesticide application. Some backpack sprayers are hand-operated and others are powered by small gasoline engines. Also known as a *knapsack sprayer*.

bacterium. a one-celled microscopic plantlike organism that lives in soil, water, or organic matter, or in the bodies of plants and animals. Some bacteria cause plant or animal diseases. (plural: *bacteria*.)

bait. a food or foodlike substance that is used to attract and often poison pest animals.

bait station. a box or similar device designed to hold poisoned bait for controlling rodents, insects, or other pests. Bait stations usually have baffles or small openings to prevent access to the bait by nontarget animals.

banding. the application of liquid or dry pesticides in bands or strips, usually to the soil, rather than over the entire area.

belly grinder. a hand-operated device used to apply pesticide granules. The device straps to the front of the operator, who then turns a crank while walking forward through the treatment area.

beneficial. helpful in some way to people, such as a beneficial plant or insect.

beneficial organism. an organism that is helpful in some way to people, such as a parasitic wasp or predatory insect.

biennial. a plant that completes part of its life cycle in one year and the remainder of its life cycle the following year.

binding. a chemical reaction in which molecules of one substance attach to molecules of another substance, forming a bond that can only be broken through another chemical reaction. Also known as *chemical binding*.

binding agent. a special type of adjuvant, similar to glue, that is designed to prevent a sprayed pesticide from washing or rubbing off of the treated surface.

bioaccumulation. the gradual buildup of certain pesticides within the tissues of living organisms after they feed on lower organisms that contain smaller amounts of these pesticides. Animals that are higher up on the food chain accumulate greater amounts of these pesticides in their tissues.

biological activity. a chemical's activity that involves the biological processes of living organisms, as opposed to physical or mechanical activity.

biological control. the action of parasites, predators, pathogens, or competitors in maintaining another organism's numerical density at a lower average than would occur in their absence. Biological control may occur naturally or it may be the result of the manipulation or introduction of biological control agents by people.

biological factors (in pesticide resistance). factors such as life cycles, life stages, physical attributes, and others that protect certain organisms from the toxic effects of pesticides.

biology. knowledge about the life habits of a plant or animal.

biotic. pertaining to living organisms, such as the influences living organisms have on all pest populations.

blacklight trap. a device that uses ultraviolet light to attract insects.

boom. an extension structure attached to a truck, tractor, or other vehicle, or held by hand, to which spray nozzles are attached.

boom applicator. a pesticide application device having multiple nozzles spaced along a boom, making it possible to spray a wide swath. Boom applicators usually are used for herbicides and for other pesticides in turf.

botanical. derived from plants or plant parts.

brand name. the registered or trade name given to a pesticide by its manufacturer or formulator. A specific pesticide may be sold under several brand names.

breakdown. the process by which chemicals, such as pesticides, decompose into other chemicals.

broadcast application. a method for applying granular pesticides by dispersing them over a wide area using a spinning disc or other mechanical device.

broadleaves. one of the major plant groups, also known as *dicots,* with net-veined leaves that usually are broader than grass leaves. Broadleaf seedlings have two seed leaves (*cotyledons*). Broadleaves include many herbaceous plants, shrubs, and trees.

broad-spectrum pesticide. a pesticide that is capable of controlling many different species or types of pests.

bryophyte. a non-flowering plant such as moss or liverwort.

buffer. an adjuvant that lowers the pH of a spray solution and, depending on its concentration, can maintain the pH within a narrow range even if acidic or alkaline materials are added to the solution.

buffer strip. a strip at the edge of a pesticide-treatment area that is left untreated in order to protect adjoining areas from pesticide hazards. The minimum buffer strip width is usually one swath width for the applicator equipment.

calibration. the process used to measure the output of pesticide application equipment so that the proper amount of pesticide can reliably be applied to a given area.

California Department of Pesticide Regulation (DPR). the state agency responsible for regulating the use of pesticides in California.

carcinogenic. having the ability to produce cancer.

cardiopulmonary resuscitation (CPR). a procedure designed to restore normal breathing and heartbeat after breathing and heartbeat have stopped.

carrier. the liquid or powdered inert substance that is combined with the active ingredient in a pesticide formulation. May also apply to the water or oil that a pesticide is mixed with prior to application.

CAS (Chemical Abstract Service Registry). a function of a unit of the American Chemical Society that indexes chemicals by various physical and chemical characteristics and assigns each chemical a registry number so that this information can be retrieved.

caustic. quality of a chemical describing its ability to burn or injure the skin, eyes, or mouth and intestinal lining.

Caution. the *signal word* used on labels of the least-toxic pesticides; these pesticides have an oral LD_{50} greater than 500 mg/kg and a dermal LD_{50} greater than 2,000 mg/kg. Also see *LD_{50}* and *signal word.*

cephalothorax. the combined head and thorax of spiders from which arise the eyes, legs, pedipalps, and mouth structures.

CERCLA (Comprehensive Environmental Response, Compensation, and Liability Act). commonly known as Superfund, this Act enacted by Congress on December 11, 1980, created a tax on the chemical and petroleum industries and provided broad federal authority to respond directly to releases or threatened releases of hazardous substances that may endanger public health or the environment.

Certificate of Good Standing. a certificate issued by the office of the California Secretary of State that shows the status of a business or corporation. This certificate is required by the Department of Pesticide Regulation as part of the application for a landscape management pest control business license or a maintenance gardener pest control business license.

Certificate of Insurance. a certificate issued by an insurance company that verifies that a business is insured. This certificate is required by the Department of Pesticide Regulation as part of the application for a landscape management pest control business license or a maintenance gardener pest control business license.

certified pesticide applicator. a person who has demonstrated through an examination process the ability to safely handle and apply highly hazardous, restricted-use pesticides.

certified private applicator. a property owner, manager, or responsible person employed by the property owner or manager who has demonstrated through an examination process the ability to safely handle and apply restricted-use pesticides on the property under his or her control.

chemical name. the official name given to a chemical compound to distinguish it from other chemical compounds.

CHEMTREC. a chemical-industry-supported organization that provides assistance and advice on pesticide emergencies. The telephone number of CHEMTREC is 1-800-424-9300.

chronic. pertaining to long duration or frequent recurrence.

chronic illness. an illness that will last for a long period. Cancer, respiratory disorders, and neurological disorders are examples of chronic illnesses that have been associated with exposure to some types of pesticides.

Class 1 Disposal Site. a disposal site for toxic and hazardous materials such as pesticides and pesticide-contaminated wastes.

Class 2 Disposal Site. a disposal site for nontoxic and nonhazardous materials such as household and commercial waste. A sanitary landfill is a Class 2 disposal site.

commercial park. a business that provides outdoor entertainment in the form of rides, tours, or other activities; commercial parks usually contain landscaped areas and turf.

common name. the recognized, nonscientific name given to plants or animals. The Weed Science Society of America and the Entomological Society of America publish lists of recognized common names. Many pesticides also have common names, separate from their brand names and chemical names.

compacted soil. a condition where soil particles are closely pressed together because of people's activities, heavy equipment, or water ponding. Compacted soil has water drainage and penetration problems.

compatible. the condition in which two or more pesticides mix without unsatisfactory chemical or physical changes.

competition. the struggle between different organisms for the same resources, such as water, light, nutrients, or space.

compressed air sprayer. a hand-held sprayer that uses a hand-operated air pump to pressurize the liquid pesticide mixture held in a sealed tank. The air pressure forces the pesticide out of the tank and through a nozzle or other orifice.

cone nozzle. a style of nozzle that produces liquid pesticide sprays in a solid cone pattern or a hollow cone pattern.

confined area. an area such as the inside of a building or greenhouse, attic, crawl space, or the hold of a ship that may have restricted air circulation and may therefore promote the buildup of toxic fumes or vapors from a pesticide application.

contact poison. a pesticide that provides control when target pests come into physical contact with it.

control agent. an organism or chemical, such as a natural enemy or a pesticide, that reduces pest populations.

convulsions. contortions of the body caused by violent, involuntary muscular contractions. Convulsions can be a symptom of pesticide poisoning.

corrosive materials. certain chemicals that react with metals or other materials. Some pesticides are corrosive, and special handling requirements are needed when using these.

cotyledon. the first leaf or pair of leaves (*cotyledons*) of a sprouted seed. Grasses (*monocots*) have a single cotyledon while broadleaved plants (*dicots*) have a pair of cotyledons.

coverage. the degree to which a pesticide is distributed over a target surface.

coverall. a one- or two-piece garment of closely woven fabric that covers the entire body except for the head, hands, and feet, and must be provided by the employer as personal protective equipment. Coveralls differ from, and should not be confused with, work clothing that an employer can require an employee to provide.

CPR. see *cardiopulmonary resuscitation.*

cross resistance. a condition where an organism that has developed resistance to one type or group of pesticides is also resistant to other similar or dissimilar pesticides even though the organism has never been exposed to those other pesticides.

cumulative effect. poisoning symptoms that only appear after several repeated doses over time, indicating that the toxic effect is building up in the system of the poisoned individual.

cuticle. the outer protective covering of plants and arthropods that helps prevent moisture loss.

Danger. the *signal word* used on labels of highly hazardous pesticides—those pesticides having an oral LD_{50} of less than 50 mg/kg or a dermal LD_{50} of less than 200 mg/kg or those having specific, serious health or environmental hazards. Also see LD_{50} and *signal word.*

deactivation. the process by which the toxic action of a pesticide is reduced or eliminated by impurities in the spray tank, the water being used for mixing, or biotic or abiotic factors in the environment.

decomposition. the process by which pesticides break down into other chemicals, usually having no further pesticidal activity. Sunlight, temperature, soil microorganisms, and other factors contribute to pesticide decomposition.

decontamination. the most important step in reducing potential injury when someone has been exposed to a pesticide. Decontamination involves thoroughly washing the exposed skin with soap and water or flushing the exposed eye with a gentle stream of running water.

defoaming agent. an adjuvant that keeps a pesticide mixture from foaming in a spray tank.

degradation. the breakdown of a pesticide into an inactive or less-active form. Environmental conditions, impurities, or microorganisms can contribute to the degradation of pesticides.

dehydration. the process of a plant or animal losing water or drying up.

deposition. the placement of pesticides on target surfaces.

deposition aid. an adjuvant that improves the ability of a pesticide spray to reach the target.

dermal. pertaining to the skin. Dermal entry is one of the major ways pesticides can enter the body and possibly cause poisoning.

dermal absorption. See *skin absorption.*

dermatitis. inflammation, itching, or irritation of the skin caused by pesticide exposure.

desiccant. a material that removes water from plants or arthropods or destroys the waxy coatings that protect these organisms from water loss; used for weeds and arthropods as well as to defoliate certain plants before harvest.

desorbed, desorption. the release of pesticide materials that previously had become absorbed into or adsorbed onto a surface.

dethatching. a process that removes or breaks up thatch in turf.

detoxify, detoxification. a process that is used to render a chemical nontoxic. Some organisms can detoxify pesticides through internal biological processes.

dicot. a member of a group of plants whose seedlings have two *cotyledons;* dicots are also known as *broadleaves* or broadleaf plants. See also *monocot.*

diluent. the inert liquid or powdered material that is combined with a pesticide's active ingredient during manufacture of a pesticide formulation. Also, the water, petroleum oil, or other liquid in which the formulated pesticide is mixed before application.

directions for use. the instructions found on a pesticide label indicating the proper procedures for mixing and application.

disease. a condition, caused by biotic or abiotic factors, that impairs some or all of the normal functions of a living organism.

dispersion. the act of spreading pesticide droplets, dusts, or granules widely over a target area.

disposal site. see *Class 1 disposal site* and *Class 2 disposal site.*

dissolve. to pass into solution.

dormant. inactive during winter or periods of cold weather.

dose. the measured quantity of a pesticide. Often the size of the dose determines the degree of effectiveness, or, in the case of poisoning of nontarget organisms, the degree of injury.

drift. the movement of pesticide dust, spray, or vapor through the air away from the application site.

drift-control agent. an adjuvant added to a liquid pesticide mixture that increases the spray droplets' size, making droplets larger and less prone to drift. Drift-control agents usually thicken the spray liquid.

dry flowable. a dry, granular pesticide formulation intended to be mixed with water for application. When combined with water, a dry flowable will be similar to a wettable powder. Dry flowable formulations are measured by volume rather than by weight.

dust. finely ground pesticide particles, sometimes combined with inert materials. Dusts are applied without being mixed with water or other liquids.

ecological. related to an approach that considers the interrelationship between living organisms and the environment.

economic damage. damage from pests to plants, animals, or other causes that results in loss of income or a reduction of value.

effective life. the period of time that an applied pesticide remains toxic enough to adequately control the pests for which it was intended. Also see *shelf life*.

efficacy. the ability of a pesticide to produce a desired effect on a target organism.

electrostatic. an electrical charge that causes a pesticide liquid or dust to be attracted to the target surface.

elytra (plural). the first pair of wings of an adult beetle; the elytra provide protection and a covering for the second pair of wings, which beetles use for flight.

emergence. the appearance of a plant through the surface of the soil.

emulsifiable concentrate. a pesticide formulation consisting of a petroleum-based liquid and emulsifiers that enable it to be mixed with water for application.

emulsifier. an adjuvant that is added to a pesticide formulation to permit petroleum-based pesticides to mix with water.

encapsulation. a process by which tiny liquid droplets or dry particles are packaged into polymer plastic capsules to slow their release into the environment and prolong the application's effectiveness. Sometimes encapsulation reduces the hazard to people mixing or applying pesticides.

endangered species. rare or unusual living organisms whose existence is threatened by people's activities, including the use of some types of pesticide.

environment. all of the living organisms and nonliving features of a defined area.

environmental contamination. the spread of pesticides away from the application site into the environment, usually with the potential to cause harm to organisms.

Environmental Protection Agency (EPA). the federal government agency responsible for regulating pesticide use in the United States.

epidermis. for vertebrates, the outer layer of skin, or for invertebrates, the cellular layer of tissue beneath the cuticle.

eradicant. a pesticide that is used to destroy a pest organism such as a fungus.

eradication. the pest management strategy that attempts to eliminate all members of a pest species from a defined area.

estivation. see *aestivation*.

evaporate, evaporation. the process by which a liquid turns into a gas or vapor.

even flat-spray nozzle. a nozzle that provides a uniform distribution of spray throughout its fan pattern.

exclusion. a pest management technique that uses physical or chemical barriers to prevent certain pests from getting into a defined area.

exposure. the unwanted contact of pesticides or pesticide residues with people, other organisms, or the environment.

extender. an adjuvant that enhances the effectiveness or effective life of a pesticide by some means, such as screening ultraviolet light, slowing down volatilization, or improving sticking qualities.

federally restricted. an EPA classification for pesticides that restricts their use to certified pesticide applicators.

fibrous. thin, long, and multibranching roots that form a dense clump.

Fictitious Business Name Statement. a statement issued by county or city authorities and required as part of the Department of Pesticide Regulation's application for a landscape management pest control business license or a maintenance gardener pest control business license if the business is operating under a name other than the name of the legal owner.

FIFRA. the Federal Insecticide, Fungicide, and Rodenticide Act. This is the federal law that regulates pesticide registration, labeling, use, and disposal in the United States.
filamentous. long and threadlike.
first aid. the immediate assistance provided to someone who has been exposed to a pesticide. First aid for pesticide exposure usually involves removal of contaminated clothing and washing of the affected area of the body to remove as much of the pesticide material as possible. First aid is not a substitute for competent medical treatment.
flaming. a nonchemical weed control method involving the use of a propane-fueled torch to burn or scorch emerged weeds.
flat-spray nozzle. a nozzle that produces a fan-shaped pattern with more droplets in the center part of the fan that at either edge, allowing for overlap of the spray pattern and providing an even application when multiple nozzles are used on a boom.
flood nozzle. a nozzle that applies large volumes of liquid under low pressure; more commonly used for liquid fertilizer application than for pesticide application.
flow rate. the amount of pesticide being expelled by a pesticide sprayer or granule applicator per unit of time.
flowable. a flowable formulation consists of finely ground particles of pesticide active ingredient mixed with a liquid, along with emulsifiers, to form a concentrated emulsion. These liquids are mixed with water for dilution prior to spraying.
foliage. the leaves of a plant.
formulation. a manufacturer's mixture of pesticide active ingredient with inert materials. Inert materials are added to improve the mixing and handling qualities of the pesticide.
frass. solid fecal material produced by insect larvae.
fruiting bodies. special structures produced by fungi that contain the spores by which the organisms reproduce.
fume. the vapor phase of some pesticide active ingredients.
fumigant. the vapor or gas form of a pesticide; used to penetrate porous surfaces for control of soil-dwelling pests or of pests in enclosed areas or storage facilities.
fumigation. the process of controlling certain pests by exposing them to an atmosphere of toxic gas inside an enclosed area or under tarped soil.
fungicide. a pesticide used for control of fungi.
fungus. a multicellular lower organism such as a mold, mildew, rust, or smut. The fungus body normally consists of filamentous strands called the mycelium and reproduces through dispersal of spores. (plural: *fungi.*)
general-use pesticide. a pesticide that has been designated for use by the public as well as by licensed or certified applicators. General-use pesticides usually have minimal hazards and do not require a permit for purchase or use.
gnathosoma. the small head structure of a mite, from which small appendages used for feeding arise.
granule, granular formation. a dry formulation of pesticide active ingredient and inert materials compressed into small, pebble-like shapes.
groundwater. fresh water trapped in aquifers beneath the surface of the soil; one of the primary sources of water for drinking, irrigation, and manufacturing.
Groundwater Protection List. in California regulations, a list of pesticides that are labeled for agricultural, outdoor institutional, or outdoor industrial use and are designated as having the potential to pollute groundwater.
ground-wheel-driven. describes a hand-operated or trailer-mounted granule applicator that uses the rotation of one of its wheels to drive the operation of its dispersal mechanism (often a spinning disc).
habitat. the place where plants or animals live and grow.
hand-operated backpack sprayer. a backpack sprayer that has a hand-operated hydraulic pump, operated by moving a lever up and down, that forces liquid pesticides through one or more nozzles. See also *powered backpack sprayer.*
hand spray gun. a special device used to apply a high-pressure stream or spray of insecticide or fungicide to trees and shrubs
handler. a person who mixes, loads, transfers, applies, or assists with the application of pesticides; who maintains, services, repairs, cleans, or handles equipment used in these activities; who works with unsealed pesticide containers; who adjusts, repairs, or removes treatment site coverings; or who incorporates pesticides into the soil.
hazard. the degree of danger to people or the environment posed by a pesticide.
hazardous materials. materials, including many pesticides, that have

been classified by regulatory agencies as being harmful to the environment or to people. Hazardous materials require special handling and must be stored and transported in accordance with regulatory mandates.

hazardous waste. a hazardous material for which there is no further use. Remains from a pesticide spill cleanup often are hazardous wastes. Hazardous wastes can be disposed of only through special hazardous material incineration or through transportation to a Class 1 disposal site.

heat stress. the potentially life-threatening overheating of the body under working conditions, usually occurring when proper preventive measures are not taken, such as drinking plenty of water, taking frequent breaks in the shade to cool down, and removing or loosening personal protective equipment during breaks. California regulations require that pesticide handlers receive training on how to recognize, avoid, and treat heat stress.

HEPA. see *high-efficiency particulate air filter.*

herbaceous plant. a plant that is herblike, usually having little or no woody tissue.

herbicide. a pesticide used for the control of weeds.

high-efficiency particulate air filter. a special filtering medium designed to remove extremely small particles from the air.

high-pressure hydraulic sprayer. spraying equipment that produces a large, powerful stream of spray mixture; usually equipped with a *hand spray gun.*

honeydew. the sweet, sticky fluid secreted by some plant-feeding insects such as aphids and scale. A coating of honeydew on leaf or fruit surfaces promotes the buildup of a fungus known as sooty mold. Ants, bees, wasps, and other insects feed on honeydew and may protect the plant-feeding insects from attack by other insects.

hopper. the bin of a granule applicator where the pesticide is placed for application; granules move from the hopper to the metering device for dispersal.

hormonal herbicide. a special type of herbicide that controls weeds by changing their growth rates or patterns.

hormone. a chemical produced in the cells of a plant or animal that produces changes in cells in another part of the organism's structure.

horticulture advisors. University of California specialists in most counties of California who serve as a resource for state residents on pest management, water management, soil management, plant nutrition, and many other issues.

hose-end sprayer. a simple application device for small areas. It consists of a glass or plastic jar holding a pesticide mixture that is sucked up by and diluted in a stream of water from a garden hose and then sprayed out through a nozzle.

host. a plant or animal species that provides sustenance for another organism.

host resistance. the ability of a host plant or animal to ward off or resist attack by pests or to be able to tolerate damage caused by pests.

hydrolysis. a chemical process that involves incorporating a water molecule into another molecule.

hygiene. as it applies to pesticide exposure, hygiene involves washing exposed body areas promptly to remove pesticide residues.

hypha. a threadlike fiber that makes up the mycelium of a fungus. (plural: *hyphae.*)

idiosoma. in mites, the joined thorax and abdomen from which the pairs of legs arise.

illegal. against the law.

impermeable. having the ability to resist penetration by a substance or object.

implantation. a method of applying certain systemic pesticides to trees to control tree pests. Capsules containing the pesticide material are implanted into the bark of the tree.

incidental. pertaining to pesticide application services that might be provided by maintenance gardeners for their clients as part of a landscape maintenance business, but not to pesticide application services performed for hire as the primary function of the business.

incompatibility. a condition in which two or more pesticides are unable to mix properly or one of the materials chemically alters the other to reduce its effectiveness or to produce undesirable effects on the target.

incompatible mixture. the result when two or more pesticides are combined and they react to make the mixture unusable.

incorporate. to move a pesticide below the surface of the soil through tilling or irrigation; to combine one pesticide with another.

inert. not having any chemical activity.

inert ingredients. all materials in the pesticide formulation other than the active ingredient. Some ingredients classified as may still be toxic or hazardous to people.

infection. the establishment of a microorganism within the tissues of a host plant or animal.

infestation. a troublesome invasion of pests within an area such as a landscaped location.

inflorescence. the reproductive shoot system of a plant that bears flowers.

ingest. to take into the body through the mouth, as by eating or drinking.

inhalation. the method of entry of pesticides through the nose or mouth into the lungs.

injection. a method for applying certain systemic pesticides to trees to control tree pests. Containers or vials hold the pesticide material that is injected into the tree at its base and the material is transported to the leaves or roots through the tree's sap.

insecticide. a pesticide used for the control of insects. Some insecticides are also labeled for control of ticks, mites, spiders, and other arthropods.

instar. the period between molts in insect larvae. Most larvae pass through several instars; these are usually given numbers such as 1st instar, 2nd instar, etc.

integrated pest management (IPM). a pest management program that uses life history information and extensive monitoring to understand a pest and its potential for causing economic damage. Control is achieved through multiple approaches including prevention, cultural practices, pesticide applications, exclusion, natural enemies, and host resistance. The goal is to achieve long-term suppression of target pests with minimal impact on nontarget organisms and the environment.

intentional misapplication. the deliberate improper use of a pesticide, such as exceeding the label rate or applying the material to a site not listed on the label.

interactive effect. interaction when two or more pesticides are mixed, producing greater or lesser toxicity to the target pests or changing the mode of action.

introduced pest. a pest that is transported from its native area to a location where it did not previously exist. Some pests are introduced accidentally while others have been introduced intentionally.

inversion. a weather phenomenon in which cool air near the ground is trapped by a layer of warmer air above. Vapors from pesticides applied during an inversion can become trapped and concentrated in the lower air layer, and then have the potential to move away from the treatment area, causing damage or injury at some other location. Also known as *temperature inversion.*

invert emulsion. an emulsion where water droplets are suspended in oil rather than the oil droplets being suspended in water.

invertebrate. any animal not having an internal skeleton or shell, such as an insect, spider, mite, worm, nematode, snail, or slug.

ion. an atom or molecule that carries a positive or negative electrical charge due to the loss or gain of electrons through a chemical reaction.

ionize. to convert a chemical's molecules into ions by dissolving the chemical in water or some other liquid.

irreversible injury. a health condition caused by certain types of exposure to some pesticides, and for which there is no medical treatment or chance of recovery.

key pest. a pest that causes serious damage to landscape, interiorscape, or turf plants unless it is managed in some way.

knapsack sprayer. see *backpack sprayer.*

knowledge expectations. the breadth of knowledge about an occupation or procedure, such as pesticide handling, that a person performing this job is expected to have. Regulations establish minimal expectations for pesticide applicators, and certification examinations test whether a person's knowledge meets these expectations.

labeling. the pesticide label and all associated materials, including supplemental labels, special local needs registration information, and manufacturer's information. The pesticide label is a legal document.

larva. the active immature form of insects that undergo metamorphosis to reach adulthood. (plural: *larvae.*)

LC_{50}. the lethal concentration of a pesticide in the air or in a body of water that will kill half of the animals in a test population. LC_{50} values are given in micrograms per milliliter of air or water (µg/ml).

LD_{50}. the lethal dose of a pesticide that will kill half of the animals in a test population. LD_{50} values are given in milligrams per kilogram of test animal body weight (mg/kg).

leaching. the process by which some pesticides move down through the soil, usually dissolved in water, with the possibility that the pesticide will reach groundwater.

lethal. capable of causing death.
lethal concentration. see *LC_{50}*.
lethal dose. see *LD_{50}*.
life stage. any one of the development stages that a living organism passes through over time. Plants and animals (especially insects) pass through several life stages during which their susceptibility or tolerance to pesticides varies.
long-term health problem. a pesticide-related illness or disease that may extend over months, years, or a lifetime.
low-pressure flat-spray nozzles. nozzles that are similar to flat-spray nozzles but can perform adequately under lower pressures. The spray droplets are larger than those of a high-pressure spray, and this reduces the potential for drift.
low-pressure sprayer. spraying equipment that is designed to function properly at a low pressure, reducing the potential for spray drift.
Material Safety Data Sheet (MSDS). an information sheet provided by a pesticide manufacturer to describe the chemical qualities, hazards, safety precautions, and emergency procedures to be followed in case of a spill, fire, or other emergency.
medical facility. a clinic, hospital, or physician's office where immediate medical care for pesticide-related illness or injury can be obtained.
mesh. the number of wires per inch in a screen, such as a screen used to filter foreign particles out of a spray solution to keep nozzles from becoming clogged. Mesh is also used to describe the size of pesticide granules, pellets, and dust particles.
metabolism. the overall chemical process that takes place in a living organism to utilize food and manage wastes, provide for growth and reproduction, and accomplish all other life functions.
metamorphosis. the changes that take place in certain types of living organisms, such as insects, as they develop from eggs through adults. Some families of insects undergo "complete metamorphosis," meaning that their young do not resemble their adults.
microencapsulated. a pesticide formulation in which particles of the active ingredient are encased in plastic capsules; the pesticide is released after application, as the capsules break down.
micron. a very small unit of measure: 1/1,000,000 of a meter, or 1/1,000 of a millimeter.
microorganism. an organism of microscopic size, such as a bacterium, virus, fungus, viroid, or mycoplasma.
mixing. the process of opening pesticide containers, weighing or measuring specified amounts, and transferring these materials into application equipment, all in accordance with instructions found on pesticide labels.
mode of action. the mechanism by which a pesticide kills or controls the target organism.
molluscicide. a pesticide used to control slugs and snails.
molting. in invertebrates such as insects and spiders, the process of shedding the outer body covering or exoskeleton. Molting usually takes place to allow the animal to grow larger.
monitoring (pest monitoring). the process of carefully watching the activities, growth, and development of pest organisms over time, often utilizing very specific, conventional procedures.
monocot. a member of a group of plants whose seedlings have a single *cotyledon;* some monocots are known as grasses. See also *dicot.*
Monthly Pesticide Use Report. a form that in California must be completed and submitted to the local agricultural commissioner's office by the tenth of the month following any month in which agricultural-use pesticides are applied in landscapes or on turf or in which pesticides are applied for hire to landscapes, turf, or interiorscapes, or incidentally to the activities of a landscape, turf, or interiorscape maintenance business.
MSDS. see *Material Safety Data Sheet.*
mutagenic. of a chemical, having the capacity to cause inheritable abnormalities in living organisms.
mycelium. the vegetative body of a fungus, consisting of a mass of slender filaments called hyphae. (plural: *mycelia.*)
native. an animal or plant that is indigenous to an area.
natural enemy. an organism that can kill a pest organism; includes predators, pathogens, parasites, and competitors.
naturescape. an area with natural landscaping that attracts and serves as a home or refuge for wildlife.
necrosis. localized death of tissue on a living organism.
negligent application. a pesticide application in which the applicator fails to exercise proper care or to follow label instructions, with the

potential result of injury to people or surrounding areas.

nematicide. a pesticide used to control nematodes.

nematode. an elongated, cylindrical, non-segmented worm. Most nematodes are microscopic; some are parasites of plants or animals.

neoprene. a synthetic rubber material used to make gloves, boots, and clothing for protection against pesticide exposure.

no observable effect level. see *NOEL*.

NOEL (no observable effect level). the NOEL is the maximum dose or exposure level of a pesticide that will produce no noticeable toxic effect on test animals.

nonpoint source pollution. pollution from pesticides or other materials that arises from their normal or accepted use over a large general area and over an extended period.

nonselective. of a pesticide, having action against many pest species rather than just a few.

nontarget organism. an animal or plant that lies within a pesticide-treated area but is not the intended target of the pesticide application.

Notice of Intent. oral or written notification to the county agricultural commissioner, as specified by the commissioner, prior to the use of a pesticide. Notice of Intent is one condition that must be met by a Restricted-Use Permit holder.

notification. see *oral notification* and *posting*.

noxious. harmful to living organisms; for example, a noxious weed.

nozzle. the mechanical device at the tip of a pesticide applicator from which the pesticide is discharged. Type and size of nozzle determine the application rate, droplet size, and spray pattern.

occasional pest. a pest that does not recur regularly, but causes damage from time to time because of changing environmental conditions or other factors.

ocular. pertaining to the eye. This is one of the routes by which pesticides can enter the body.

off-site movement. any movement of a pesticide from the location where it was applied. Off-site movement occurs through drift, volatilization, percolation, water runoff, pruning, and blowing dust, as well as transportation on organisms or equipment.

operator identification number. the number assigned to an operator of agricultural or other property, as defined in California regulations, where agricultural-use pesticides will be applied. Operator identification numbers are used in the pesticide use reporting process.

oral. pertaining to the mouth. This is one of the routes by which pesticides can enter the body.

oral notification. the spoken notification of workers about pesticide applications that take place on property where they are employed.

organic matter. leaves, stems, bark, and other plant material that breaks down and contributes to the fertility and structure of soil where plants grow.

organism. any living thing.

ornamentals. cultivated plants that are grown for purposes other than production of food or fiber.

OSHA (Occupational Safety and Health Administration). the U.S. governmental agency that oversees and enforces laws designed to protect workers at the workplace. The equivalent agency in California state government is Cal-OSHA.

output rate. the volume of pesticide spray produced by a piece of application equipment over time, usually expressed in gallons per minute (gpm).

panicle. a flower head of a plant in which the lateral branches of the *raceme* are branched.

parasite. a plant or animal that derives all of its nutrients from another organism. Many parasites attach themselves to a host or invade the host's tissues. Parasitism may result in injury or death to the host.

parts per billion (ppb). one of the ways a pesticide can be diluted for application; a solution of 1 ppb contains one part of a designated chemical for every one billion equivalent parts of water.

parts per million (ppm). one of the ways a pesticide can be diluted for application; a solution of 1 ppm contains one part of the pesticide formulation for every one million equivalent parts of water.

pathogen. a microorganism that causes a disease.

pellet. a pesticide formulation consisting of the dry active ingredient and various inert materials pressed into uniform-sized granules.

penetrate. to pass through a surface such as skin, protective clothing, a plant's cuticle, or an insect's cuticle. Also, of an applied spray: to pass through dense foliage.

percentage. one of the ways a pesticide can be diluted for application; solutions are made that contain a certain percentage (usually by weight) of the pesticide formulation or its active ingredient.

percolation. the process by which water flows downward through permeable soil. During percolation, water may dissolve or leach pesticides and other chemicals out of the soil and carry them downward, sometimes as far as the *aquifer.*

perennial. a plant that lives longer than two years, or in some cases, indefinitely. Some perennial plants lose their leaves and become dormant during winter; others may die back and resprout from underground root or stem structures each year. The evergreens are perennial plants that do not die back or become dormant.

persistent pesticide. a pesticide that remains active in the environment for long periods because it is not easily broken down by microorganisms or environmental factors.

personal hygiene. see *hygiene.*

personal protective equipment (PPE). devices and garments that protect handlers from exposure to pesticides. These include coveralls, eye protection, gloves and boots, respirators, aprons, and hats.

pest. an organism that interferes with the availability, quality, or value of a crop, plant, or other managed resource or environment.

Pest Control Business License. the license that a business must obtain from the California Department of Pesticide Regulation before it can engage in pest control for hire.

Pest Notes. a resource on the internet developed by the UC Statewide Integrated Pest Management Program, providing up-to-date information on managing pests in landscape, turf, interiorscape, and other settings. Pest notes can be viewed at http://www.ipm.ucdavis .edu/PMG/selectnewpest.home.html.

pest resurgence. see *resurgence.*

pesticide. any substance or mixture of substances intended for preventing, destroying, repelling, or mitigating any insect, rodent, nematode, fungus, plant, or any other living thing that is recognized as a pest; also, any substance or mixture of substances intended for use as a plant growth regulator, defoliant, or desiccant.

pesticide container. a pesticide material's original packaging, including the pesticide label.

pesticide deposition. see *deposition.*

pesticide fate. what happens to a pesticide after it has been released into the environment.

pesticide formulation. the pesticide material as it comes in its original container, consisting of the active ingredient blended with inert materials.

pesticide handler. see *handler.*

pesticide resistance. genetic qualities of a pest population that enable individuals to resist the effects of certain types of pesticide that are toxic to other, less-resistant members of that species.

Pesticide Safety Information Series. a series of informational sheets developed and distributed by the California Department of Pesticide Regulation that pertain to pesticide handling, personal protective equipment, emergency first aid, medical supervision, and other pesticide-related topics.

pesticide use hazard. the potential for a pesticide to cause injury or damage during handling or application.

pesticide use record. a record of the pesticide applications made to a specific location.

pH. a measure of the concentration of hydrogen ions in a solution. As the number of hydrogen ions increases, the pH reading gets lower and the solution becomes more acidic.

pheromone. a chemical produced by an animal to attract or otherwise affect the behavior of other animals of the same species.

photosynthesis. the process by which a plant converts sunlight, minerals, and carbon dioxide into its food.

physiological. pertaining to the functions and activities of living tissues.

phytotoxic. injurious to plants.

plant growth regulator (PGR). a pesticide used to regulate or alter the normal growth of plants or the development of plant parts.

plant kingdom. one of the major groups of living organisms, another being the animal kingdom.

point source pollution. pollution of the soil or groundwater caused by spilling or dumping quantities of a toxic material in one identifiable location.

poison. a substance, such as certain pesticides, that will cause serious injury or death if small amounts are ingested, inhaled, or absorbed through the skin or eyes.

post-application cleanup. the washing of application equipment, personal protective equipment, and other items used during a pesticide application in order to remove pesticide residues.

postemergent. of an herbicide, applied after emergence of a specified weed or crop.

posting. the placing of signs around an area to inform workers and the public that the area has been treated with a pesticide.

potency. the toxicity of a pesticide.

powder. a fine-ground dust containing pesticide active ingredient and inert materials. This powder is mixed with water and then applied as a liquid spray.

powered application equipment. application equipment that uses a gasoline or electric motor to pressurize the spray liquid or to propel granules or dusts for application.

powered backpack sprayer. a backpack sprayer that uses a gasoline motor to pump liquid to the nozzles, in contrast to a *hand-operated backpack sprayer.*

ppb. see *parts per billion.*

PPE. see *personal protective equipment.*

ppm. see *parts per million.*

precautionary statements. the section on a pesticide label that lists human and environmental hazards. Personal protective equipment requirements are also listed there, along with first aid instructions and information for physicians.

precipitation. the settling of solid particles out of a solution, such as a formulated pesticide in a spray tank.

predaceous. having the habit of hunting and eating other animals.

predacide. a pesticide used for control of predaceous mammals such as coyotes.

preemergent. of an herbicide, applied before weed emergence to control the weeds as they sprout from seeds and before they push through the soil surface.

pressure. the amount of force (measured in pounds per square inch [psi]) that the application equipment pump applies to the liquid pesticide mixture in order to force it through the applicator's nozzles.

pressure gauge. an instrument on liquid pesticide application equipment that measures the pressure in the system.

preventive pest control. steps taken to prevent pest damage or infestation, including such methods as planting pest-resistant plant varieties.

private applicator. an individual who applies pesticides on agricultural or certain other property under his or her control and for his or her own benefit or needs.

private applicator certification. see *certified private applicator.*

propellant. a material, such as compressed air or gas, used to propel a spray liquid or dust onto target surfaces.

protectant. a pesticide that provides a chemical barrier against pest attack.

protective clothing. garments of personal protective equipment that cover the body, including the arms and legs.

psi. pounds per square inch, a measure of pressure.

pupa. in insects having complete metamorphosis, an individual in the resting life stage between larval and adult forms. (plural: *pupae.*)

pyrethroid. a synthetic pesticide that mimics pyrethrin, which is a botanical pesticide derived from certain species of chrysanthemum flowers.

Qualified Applicator Certificate (QAC). the credential required for individuals applying any pesticides on behalf of a landscape or interiorscape pest management business or as part of a maintenance gardener business and for individuals applying or supervising the application of restricted materials as part of their employment on property owned or managed by their employer. To obtain a QAC, an individual must pass the required examinations administered by the California Department of Pesticide Regulation.

Qualified Applicator License (QAL). the credential required for individuals supervising the operations of a landscape or interiorscape pest management business. To obtain a QAL, an individual must pass the required examinations administered by the California Department of Pesticide Regulation.

raceme. a flower stalk with one central stem, with or without a terminal flower, and bearing lateral flowers or small bunches of flowers.

rate. the quantity or volume of liquid spray, dust, or granules that is applied to an area over a specified period.

RCRA (Resource Conservation and Recovery Act). a 1976 Act gives the U.S. EPA the authority to control hazardous wastes from "cradle to grave." This includes the generation, transportation, treatment, storage, and disposal of hazardous wastes. RCRA also sets forth a framework for the management of nonhazardous wastes.

ready-to-use (RTU). a special pesticide formulation type that requires no dilution or other type of mixing before use. RTU pesticides are often packaged in aerosol or trigger pump containers for home and garden use.

recombination. an occurrence in which a pesticide breaks down and then combines with other chemicals in the environment to produce a different compound than what was originally applied.

recommendation. a written document, properly called a "written recommendation," prepared by a licensed Pest Control Adviser that prescribes the use of a specific pesticide or other pest control method.

regularly handle. to handle pesticides during any part of the day for more than six calendar days in any 30-consecutive-day qualifying period that begins on the first day of handling.

regulations. the guidelines or working rules that a regulatory agency uses to carry out and enforce laws.

REI. see *restricted-entry interval.*

repellent. a pesticide used to keep target pests away from a treated area by saturating the area with an odor that is disagreeable to the pest.

rescue breathing. also known as *artificial respiration.* Mouth-to-mouth resuscitation administered to assist or restore the breathing of a person overcome by pesticides. Rescue breathing is administered if the victim has a pulse. See also *cardiopulmonary resuscitation.*

reservoir. a population of pests within a local area. Also an organism harboring plant or animal pathogens.

residual action. the pesticidal action of a material after it has been applied. Most pesticide compounds will remain active for from several hours to several weeks or even months after being applied.

residue. traces of pesticide that remain on treated surfaces after a pesticide application.

resistance. see *pesticide resistance* or *host resistance.*

respirator or respiratory equipment. a device that filters out pesticide dusts, mists, and vapors to protect the wearer from respiratory exposure during mixing, loading, and application, or when entering a treated area before expiration of the *restricted-entry interval.*

restricted materials. see *restricted-use pesticide.*

restricted materials permit. see *restricted-use permit.*

restricted-entry interval (REI). a period of time that must elapse after the application of a pesticide before it is safe to allow people into the treated area without requiring that they wear personal protective equipment.

restricted-use. see *restricted-use pesticide.*

restricted-use permit. a permit, issued by a county agricultural commissioner office, that enables an individual to possess and apply a restricted-use pesticide. Restricted-use permits can only be issued to certified applicators.

restricted-use pesticide. a highly hazardous pesticide that can only be possessed or used by a commercial applicator who has a valid Qualified Pesticide Applicator license or certificate or by a private applicator who has passed a written exam administered by the local agricultural commissioner, thereby demonstrating an understanding of the proper methods of handling, using, and disposing of these materials.

restrictive statement. a statement on a pesticide label that restricts the use of that pesticide to specific areas or to use by designated individuals.

resurgence. the sudden increase of a pest population after some event, such as a pesticide application.

reversible injury. a pesticide-related injury or illness that can be reversed through medical intervention and/or the body's healing process.

revolutions per minute (rpm). a measurement of the rotation speed of the spinning disk in an applicator for dry pesticide formulations.

rhizome. an underground stem of certain types of plants, often capable of growing roots at its nodes to sprout new plants.

rinsate. liquid that has been used for rinsing pesticide containers or spray equipment.

rodenticide. a pesticide used for control of rats, mice, gophers, squirrels, and other rodents.

rope wick applicator. see *wick applicator.*

route of exposure. the path by which a pesticide gets onto or into the body. The four routes of exposure are dermal (on or through the skin), ocular (on or in the eyes), respiratory (into the lungs), and oral (through swallowing).

rpm. see *revolutions per minute.*

RTU. see *ready-to-use.*

ruffling. an application technique in which spray is applied with oscillating motion or an air blast so the droplets will contact all of the target plant's leaf surfaces.

runoff. the liquid spray material that drips from the foliage of treated plants or from other treated surfaces. Also, the rainwater or irrigation water that leaves an area: this water may contain trace amounts of pesticide.

russeting. a rough, brownish area that appears on fruit and other plant parts, often as a result of infection by a pathogen, insect activity, or environmental conditions.

saprophyte. an organism that lives on dead or decaying organic matter.

sapwood. the outer layer of living tissue under the bark of a tree.

SARA Title III (the Emergency Planning and Community Right-to-Know Act). a law that establishes requirements for federal, state, and local governments, Indian tribes, and industry with regard to emergency planning and "Community Right-to-Know" reporting on hazardous and toxic chemicals. The Community Right-to-Know provisions help increase the public's knowledge of and access to information on chemicals at individual facilities, their uses, and releases into the environment. States and communities, working with these facilities, can use the information to improve chemical safety and protect public health and the environment.

School Site Pesticide Use Report. an annual report that school districts are required to file with the California Department of Pesticide Regulation by January 30 of each year, listing pesticides that have been applied to school sites.

scutellum. a triangular or shield-like plate that can be seen from above on certain insects and is located between the wings at their junction with the thorax. The scutellum is very prominent in most true bugs.

secondary pest. an organism that becomes a serious pest only after a natural enemy, competitor, or primary pest has been eliminated through some type of pest control method.

selective pesticide. a pesticide that has a mode of action against only a single pest species or a small number of pest species.

selectivity. with pesticides, the degree to which a particular pesticide will kill some organisms without harming others. Phenoxy herbicides, for example, exhibit selectivity in that they control broadleaf plants but not grasses.

self-contained breathing apparatus (SCBA). see *supplied-air respirator.*

service container. any container other than the original pesticide container that is designed to hold concentrated or diluted pesticide mixtures, including the sprayer tank.

shelf life. the maximum period of time that a pesticide can remain in storage before it loses some of its effectiveness. Also see *effective life.*

shingling. the clumping or sticking together of plant foliage caused by the force of a liquid spray. Shingling prevents spray droplets from reaching all surfaces of the foliage and may result in poor pest control.

sight gauge. a device on a pesticide sprayer or configuration of the spray tank that permits the operator to view the level of liquid in the tank.

signal word. one of three words (*Danger, Warning, Caution*) found on every pesticide label to indicate the relative hazard of the chemical. See also *toxicity category.*

site. the area where pesticides are applied for control of a pest.

site identification number. the number assigned to a location where pesticides are to be applied.

site of action. the location within the tissues of the target organism where a pesticide acts.

skin absorption. the passage of pesticides through the skin into the bloodstream or other organs of the body. Also called *dermal absorption.*

skull and crossbones. the symbol on the label of a pesticide that is highly poisonous. This symbol is always accompanied by the signal word *Danger* and the word "poison."

soil mobility. a variable characteristic of a pesticide, based on its chemical nature. Highly mobile pesticides leach rapidly through the soil and may contaminate groundwater. Immobile pesticides (those with low soil mobility) remain tightly attached to soil particles and are resistant to leaching.

soil profile. the characteristics and differences of a soil at a particular depth.

solarization. a control method for weeds, weed seeds, nematodes, and other soil-inhabiting pests that does not use pesticides. Moist soil is covered with clear plastic film during periods of hot weather. Heat from the sun is trapped beneath the film, raising the soil temperature high enough to destroy seeds and kill living organisms.

solid-stream nozzle. a special nozzle used with a hand spray gun to produce a high-pressure stream of spray for reaching tall shrubs and trees.

soluble. able to dissolve completely in a liquid.

soluble powder. a pesticide formulation in which the active ingredient and all inert ingredients completely dissolve in water to form a true solution.

solution. a liquid that contains dissolved substances, such as a soluble pesticide.

solvent. a liquid capable of dissolving certain chemicals.

sorptive dust. a fine powder used to destroy arthropods by removing the protective wax coating that prevents them from dehydrating.

source reduction. a sanitation practice that removes conditions that attract pests or are necessary for pest survival, including food, water, and shelter.

speed of travel. the speed at which the operator moves a piece of pesticide application equipment through the area being treated. One must calculate the speed of travel in order to calibrate the pesticide application equipment and achieve an accurate application dosage.

spike. a rigid tube or rod of formulated pesticide (or fertilizer) that can be inserted into the planting medium to slowly release the active ingredient for uptake into the plant for control of insects or mites.

spore. a reproductive structure produced by some plants, fungi, and microorganisms that is resistant to environmental influences.

spot treatment. a method for applying pesticides only in small, localized areas where pests congregate, rather than to a larger, general area.

spray nozzle. see *nozzle*.

spreader. an adjuvant that lowers the surface tension of treated surfaces so that a pesticide can be more readily absorbed.

statement of practical treatment. a section of the pesticide label that provides information on treating people who have been exposed to the pesticide. This includes emergency first aid information.

statement of use classification. a special statement found on labels of some highly hazardous pesticides, indicating that their use is restricted to people who are qualified to use them through a certification process.

sticker. an adjuvant used to prevent pesticides from being washed or abraded off of treated surfaces.

stippling. small, colorless dots that appear on the top or underside of plants infested with mites and certain insects. The dots are plant cells the contents of which have been removed as the pests have fed on the tissue.

stolon. an aboveground runner stem found in some plant species.

stomach poison. a pesticide that kills target animals that ingest it.

sublethal dose. a pesticide dose insufficient to cause death in the exposed organism.

summer annuals. plants (or weeds) whose seeds sprout in the spring, grow and produce seeds, and then die during the summer or fall.

supplemental label. additional instructions and information not found on the pesticide label because the physical label on the container is too small, but legally considered to be part of the pesticide's labeling.

supplied-air respirator. a tight-fitting face mask that is connected by hose either to an air supply such as a tank worn on the back of the person who is using the respirator or to an external air supply. A supplied-air respirator permits a person to enter oxygen-deficient areas or areas where there are highly toxic pesticide vapors. Also known as *self-contained breathing apparatus (SCBA)*.

suppression. the pest management strategy that attempts to reduce pest numbers to a tolerable level.

surface-active agent. see *surfactant*.

surface coverage. the degree to which a spray or dust covers the surface of leaves or other targets for treatment.

surface water. water contained in lakes or ponds or flowing in streams, rivers, and canals.

surfactant. a surface-active agent; an adjuvant used to improve a pesticide's ability to stick to and be absorbed by the target surface.

susceptible life stage. the life stage of a pest organism that is most susceptible to the pesticide used to control it. In general, insects are most susceptible during the larval or juvenile stage and weeds are most susceptible during the seedling stage.

suspension. fine particles of solid material distributed evenly throughout a liquid such as water or oil.

swath. the area covered by one pass of the pesticide application equipment.

swath width. the width of the area covered by spray droplets or granules as the application equipment moves through. You must measure the swath width in order to calibrate application equipment.

symptom. any abnormal condition that is caused by exposure to a pesticide and can be seen or felt or can be detected by examination or laboratory testing.

systemic pesticide. a pesticide that is taken up by a plant or domesticated animal and moves, after application, to other tissues in the plant or animal.

tank capacity. the largest volume of liquid that can be put into the tank of a piece of pesticide application equipment.

tank mix. a mixture of pesticides or

of fertilizers and pesticides that are applied together in a single application.

target. either the pest that is being controlled or surfaces within an infested area that the pest will contact.

target pest. see *target*.

target plant. see *target*.

temperature inversion. see *inversion*.

teratogenic. a chemical that is capable of causing non-inheritable birth abnormalities.

thatch. the leaves and other plant debris that accumulate beneath turfgrass plants at the surface of the soil.

thickener. an adjuvant that increases the viscosity of the spray solution so that larger droplets are formed by the applicator nozzles. Thickeners are used to control drift.

threshold limit value (TLV). the highest airborne concentration of a pesticide (in *parts per million*) that will produce no adverse effects over time.

topdressing. sand, fertilizer, or other materials that are spread over turfgrass following aeration to quicken healing or growth.

toxicant. a substance that, at a sufficient dose, will cause harm to a living organism.

toxicity. a pesticide's potential for poisoning an exposed organism.

toxicity category. the three classifications of pesticides that indicate the approximate level of hazard. The three categories are indicated by the signal words *Danger, Warning,* and *Caution*.

toxicity testing. a process in which known dosages of a pesticide are given to groups of test animals and the animals' responses are observed.

toxicology. the study of toxic substances and their effects on living organisms.

training record. a document signed by the trainer, employer, and trainee and used to record the dates and types of pesticide safety training that a trainee receives.

translocate, translocation. the movement of pesticides from one location to another within the tissues of a plant.

travel speed. the speed at which application equipment covers a given distance during a pesticide application, usually measured in miles per hour or feet per minute.

treated surface. the surface of plants, soil, or other items that were contacted with pesticide spray, dust, or granules for control of pests.

treatment area. see *site*.

trigger pump sprayer. a simple plastic container that holds a pesticide mixture, where the applicator expels the pesticide by squeezing a lever (trigger) while pointing the nozzle toward the target plant. Trigger pump sprayers are useful for interiorscape applications.

triple rinse. the process of partially filling an empty pesticide container, replacing the lid, shaking the container, then emptying its contents into the spray tank. This procedure is performed three times to ensure that most of the pesticide residue is removed. Triple-rinsed containers can be accepted for special pesticide container recycling or taken to a Class 2 disposal site.

TSCA (Toxic Substances Control Act). a federal law enacted to provide information about all chemicals and to control the production of new chemicals that might present an unreasonable risk of injury to human health or to the environment. TSCA authorizes U.S. EPA to require testing of old and new chemical substances. TSCA also give EPA the authority to regulate the manufacture, processing, import, and use of chemicals. Because TSCA gives EPA broad powers, the law covers virtually all manufactured and natural chemicals.

tuber. an enlarged, fleshy underground stem with buds that are capable of producing new plants.

umbel. a flower head in which all of the flower stalks originate at or near one point.

unregistered site. any site, such as a landscape area or interiorscape, that is not listed on the pesticide label. The legal use of a pesticide requires that it be applied only to listed or registered crops or sites.

use restrictions. special restrictions included on the pesticide label or incorporated into state or local regulations that specify how, when, or where a specific pesticide may be used.

vaporize. to transform from a spray of droplets to a foglike vapor or gas.

vertebrate. an animal that has an internal skeleton and segmented spine, such as a fish, bird, reptile, or mammal.

virus. a very small organism that multiplies in living cells and often is capable of producing disease symptoms in some plants and animals.

viscosity. a physical property of a fluid that affects its flowability. A more-viscous fluid flows less easily and produces larger spray droplets than a less-viscous fluid.

volatile. able to pass readily from liquid

or solid form into a gaseous form at low temperatures.

volatilization. the process by which a pesticide liquid or solid passes into a gaseous form.

Warning. the signal word used on labels of pesticides considered moderately toxic or hazardous. Pesticides that are assigned the signal word *Warning* based on toxicity usually have an oral LD_{50} between 50 and 500 mg/kg and a dermal LD_{50} between 200 and 2,000 mg/kg. Also see *LD_{50}* and *signal word.*

water-dispersible granules. a special granular formulation that releases the pesticide's active ingredient when water (whether irrigation or rainfall) contacts the ground surface where the granules have been applied.

watershed. an area of land that drains its surface water into a defined watercourse or body of water.

water-soluble. of a liquid or solid, having the ability to dissolve completely in water.

water-soluble concentrate. a liquid pesticide formulation that dissolves in water to form a true solution.

wettable powder. a type of pesticide formulation consisting of an active ingredient that will not dissolve in water, combined with mineral clay and other inert ingredients and ground into a fine powder.

wetting agent. an adjuvant used in pesticide mixtures to lower the surface tension of spray droplets, enabling them to come in close contact and spread out over target surfaces, especially those that feature fine hairs or waxy layers.

wick applicator. a special device for applying an herbicide to the surfaces of weed plants. A rope wick or felt pad becomes saturated with the herbicide mixture and this herbicide transfers to the target plant surfaces when the device is passed over the plant. An herbicide applied in this way is not subject to drift.

winter annuals. plants (weeds) whose seeds sprout in the fall, grow over the winter, and produce seeds and die before summer.

woody plant. a plant that retains some living woody material at or above ground level through the non-growing season.

work clothing. garments such as long-sleeved shirts, short-sleeved shirts, long pants, short pants, shoes, and socks. Work clothing is not considered to be personal protective equipment, although pesticide product labeling or regulations may require that specific work clothing be worn during some activities. Work clothing differs from and should not be confused with a coverall. While coveralls must be provided by the employer, an employer can require that an employee provide his or her own work clothing. Short-sleeved shirts and short pants are only considered acceptable work clothing under conditions expressly permitted by pesticide product labeling.

xylem. the tissue that carries water and nutrients up the roots and stems of plants.

Further Reading

Abiotic Disorders of Landscape Plants. 2003. L. R Costello. Oakland: Univ. Calif. Agric. Nat. Res. Publ. 3420.

IPM for Floriculture and Nurseries. 2001. S. H. Dreistadt. UC Statewide Integrated Pest Management Project. Oakland: Univ. Calif. Agric. Nat. Res. Publ. 3402.

IPM in Practice Principles and Methods of Integrated Pest Management. 2001. M. L. Flint and P. Gouveia. Oakland: Univ. Calif. Agric. Nat. Res. Publ. 3418.

Laws and Regulations Study Guide. 2001. Sacramento: State of California Department of Pesticide Regulation. Publication Number PML LRSG 1.

Natural Enemies Handbook: The Illustrated Guide to Biological Pest Control. 1998. M. L. Flint and S. H. Dreistadt. UC Statewide Integrated Pest Management Project. Oakland: Univ. Calif. Agric. Nat. Res. Publ. 3386.

Pests of Landscape Trees and Shrubs. 2004. S. H. Dreistadt. UC Statewide Integrated Pest Management Project. Oakland: Univ. Calif. Agric. Nat. Res. Publ. 3359.

The Safe and Effective Use of Pesticides. 2000. P. O'Connor-Marer. UC Statewide Integrated Pest Management Project. Oakland: Univ. Calif. Agric. Nat. Res. Publ. 3324.

UC Guide to Healthy Lawns. Available online, http://www.ipm.ucdavis.edu/TOOLS/TURF/

Weeds of the West. 2002. L. Burrill, D. Cudney, S. Dewey, R. Lee, B. Nelson, R. Parker, and R. Whitson, T. Western Society of Weed Science.

Wildlife Pest Control around Gardens and Homes. 2006. T. P. Salmon, T. A. Whisson, and R. A. Marsh. Oakland: Univ. Calif. Agric. Nat. Res. Publ. 21385.

Pest Notes

Pest Notes cover numerous topics, which are sampled below. All University of California *Pest Note* publications are available for free downloading from the UC Statewide IPM Program Web site, http://www.ipm.ucdavis.edu.

Pest Notes: Ants. 2005. Publication 7411.

Pest Notes: Aphids. 2000. Publication 7404.

Pest Notes: Bermudagrass. 2002. Publication 7453.

Pest Notes: California Ground Squirrel. 2002. Publication 7438.

Pest Notes: Carpenter Bees. 2004. Publication 7417.
Pest Notes: Cliff Swallows. 2005. Publication 7482.
Pest Notes: Cockroaches. 2000. Publication 7469.
Pest Notes: Dandelions. 1999. Publication 7467.
Pest Notes: Flies. 2004. Publication 7457.
Pest Notes: Lawn Diseases: Prevention and Management. 2002. Publication 7497.
Pest Notes: Moles. 2004. Publication 74115.
Pest Notes: Mosquitoes. 1998. Publication 7451.
Pest Notes: Pocket Gophers. 2002. Publication 7433.
Pest Notes: Powdery Mildew on Ornamentals. 2001. Publication 7493.
Pest Notes: Rats. 2003. Publication 74106.
Pest Notes: Spiders. 2000. Publication 7442.
Pest Notes: Weed Management in Landscapes. 2006. Publication 7441.
Pest Notes: Weed Management in Lawns. 2004. Publication 74113.
Pest Notes: Wood Decay Fungi in Landscape Trees. 2003. Publication 74109.
Pest Notes: Yellowjackets and other Social Wasps. 2001. Publication 7450.

Index